Molecular Biology
Concepts for Inquiry

The Exploration Workbook

JENNIFER A. HACKETT

Molecular Biology Concepts for Inquiry: The Exploration Workbook by Jennifer A. Hackett
Illustrator: Jennifer A. Hackett

Please visit the author's website for access to supporting curricular materials, to contact the author, or to share your experiences learning or teaching molecular biology. Questions from readers are welcomed. While effort has been made to ensure the accuracy of the material in this work, the author encourages readers to offer any suggestions that could correct or augment the content or clarity of future editions. The author also encourages readers to please report broken URLs. An updated list of URLs referenced in this book will be available on the author's website:
https://hackettmolecularbiology.blogspot.com/

On the cover: *Eco*RI restriction enzyme bound to its DNA recognition site. *Eco*RI is a dimer of two identical polypeptide monomers, green and blue. The DNA recognition site, G/AATTC, is palindromic and one strand is cut by each monomer in the *Eco*RI dimer. Image drawn by Jennifer Hackett using Jmol from the structure PDB ID: 1ERI, published in Kim, Y.C., Grable, J.C., Love, R., Greene, P.J., Rosenberg, J.M. Refinement of Eco RI endonuclease crystal structure: a revised protein chain tracing. *Science* **249,** 1307-1309 (1990).

ISBN-13: 9781095827925

Copyright © 2019 by Jennifer A. Hackett
All rights reserved.

Related Works

Molecular Biology: Concepts for Inquiry
Molecular Biology Concepts for Inquiry: A Guide to Inquiry

To my students and teachers.

About the Author

Jennifer A. Hackett received B.A. degrees in Biology and Chemistry from DePauw University where she worked with Chester S. Fornari and was awarded The Albert E. Reynolds Prize in Biology in 1998. In 2003, she received her Ph.D. in Human Genetics and Molecular Biology from Johns Hopkins University School of Medicine where she studied the role of telomere dysfunction in genome instability with Carol W. Greider and was awarded The Nupur Dinesh Thekdi Young Investigator Research Award. In 2003, she was awarded The Harold Weintraub Graduate Student Award from Fred Hutchinson Cancer Research Center. She completed a postdoctoral fellowship in 2006 with Stephen J. Elledge at Harvard Medical School where she contributed to the development of whole-genome shRNA screens to interrogate human diseases. While in Boston, she volunteered in programs for children through Boston Cares, motivating her to join the New York City Teaching Fellows and begin teaching high school science in 2006. As a Teaching Fellow, she received a Masters in Education from Lehman College while teaching in the New York City Public Schools. In 2008, she began teaching at an independent school in New York City, including teaching Biology, Chemistry, and a college-level Advanced Molecular Biology course that she designed in 2011. In addition to traditional teaching, she has mentored SMART Teams (protein modeling through Milwaukee School of Engineering) and an iGEM team. She has specialized in teaching science through inquiry, receiving formal training from BSCS in their 5E Inquiry method. She was recruited by BSCS to collaborate on the creation of the BSCS/NIH curriculum supplements, *Rare Diseases and Scientific Inquiry* (2011) and *Allergies and Scientific Inquiry* (2017). The first draft of the textbook, *Molecular Biology: Concepts for Inquiry,* was written during the 2011-2012 school year to facilitate the incorporation of inquiry instruction into her Advanced Molecular Biology course. Since 2012, she and her students have collaborated with Jef Boeke at Johns Hopkins University School of Medicine and NYU Langone Medical Center on research connected to Sc2.0, the Synthetic Yeast Genome Project.

Acknowledgments

My students' insatiable intellectual curiosity is the reason that this textbook had to be written. In addition, this textbook couldn't be in the form that it is today without my students. Their feedback has been key to making revisions and additions. Special thanks to Julia Masters for creating several high-quality hand-drawn illustrations for the textbook as a Senior Project in 2012. I am very grateful to Julia for her time and generosity.

Thank you to Peggy Brueggemann, my science teacher for three years at The Summit Country Day School, for providing an example of great inquiry teaching. In her 9th grade Introduction to Physical Science class, science was never presented as a set of facts and we learned through experimentation how to think like scientists. In her Biology and AP Biology courses, I learned to approach biology starting from a strong conceptual framework. She accomplished this by consciously teaching with the BSCS inquiry method. I will always be grateful for her guidance and support.

Thank you to my college mentor, Chet Fornari, whose support was invaluable and whose triplet repeat research sparked my continuing fascination with the dynamics of DNA.

Thank you to Jef Boeke for loaning me a lab bench and supplies and for including my high school students in his efforts develop educational research opportunities.

Thank you to my postdoc advisor, Steve Elledge, for the opportunity to work in his lab and for his generous support when I decided to make teaching my next step.

Thank you especially to my graduate advisor, Carol Greider, not only for facilitating a fantastic thesis experience, but also for her continued mentorship. She created an environment where graduate students could drive research and I appreciate her ongoing commitment to the support of basic science research.

Preface

In 2011, I was given the opportunity to design a new course for my high school students, which was requested to be biotechnology at the college level. As a college-level course I knew that this course would need to delve deeply into molecular biology rather than simply being a survey of fun techniques in biotechnology. I was excited to have the opportunity to actually introduce students to the concepts underlying my own research experience and the deep understanding of molecular biology I was lucky enough to be introduced to during graduate school at Johns Hopkins. I had also recently received formal training in the BSCS 5E Inquiry method for teaching science, which focuses on having students develop conceptual understanding by engaging in scientific inquiry. My best science teachers had used this approach and I definitely wanted to teach my new course through inquiry. To do so, I needed to find an appropriate textbook that I could use to replace direct lecturing so that I would have the class time needed to do lots of experiments and inquiry. I wanted to find a textbook that taught the fundamental concepts without assuming prior college-level biology and chemistry coursework. I also needed a textbook that was a reasonable length for my time-pressed students to actually read. At the time, I couldn't find an appropriate textbook, so I wrote *Molecular Biology: Concepts for Inquiry*. The inspiration for the textbook and these supporting materials comes largely from the excellent education I received as a PhD student in the Human Genetics and Molecular Biology program at Johns Hopkins University School of Medicine in Carol Greider's lab and as a postdoctoral fellow at Harvard Medical School in Steve Elledge's lab. In graduate school, in addition to the coursework and my lab work, the collaborative nature of the researchers in the Human Genetics program and in the Molecular Biology and Genetics department created a dynamic environment that interrogated a common conceptual framework that I wanted to share with students. Additionally, teaching chemistry had given me a much richer understanding of the concepts underlying biochemistry and I wanted to help my students make those connections as well.

How to Use This Inquiry Workbook

This workbook is intended to serve as a companion to the textbook *Molecular Biology: Concepts for Inquiry*. It contains exercises for individual and group inquiry explorations that can be used in a course incorporating the BSCS "5E" Inquiry method of learning science. In inquiry, students build understanding and reasoning skills by *engaging* with an idea through an activity that elicits preconceptions and misconceptions, *exploring* evidence through logic in a class activity to construct an understanding of concepts and eliminate misconceptions, *explaining* their understanding in the context of an explanation of what is known, *elaborating* on their understanding by applying it to new situations, and *evaluating* their understanding through self-assessment or other assessments. Research has shown that this is a highly effective method of learning. As a teacher, my experience has been that students learn the best when they are most actively engaged in the process of developing their own understanding instead of passively waiting for teachers or other classmates to provide answers. The 5E method effectively provides opportunities for thoughtful, active engagement in learning.

I assign textbook readings (*explaining*) to substitute for direct lecture and they are accompanied by online self-assessment reading comprehension quizzes (*evaluating*) and more open-ended questions that can serve as the basis for later class discussions. Class time is a mixture of conducting experiments (*engaging, exploring, elaborating*), completing the class activities in this workbook (*engaging, exploring, elaborating*), and conversational full-class discussions of experiments, class activities, select concepts in the textbook, and discussion questions in this workbook (*explaining, elaborating*). Questions that I have previously used for tests are included in this workbook as self-assessments (*evaluating*) and answers are provided. A subset of the class activities focuses on pre- or post-experiment analyses that could stand alone or could be used as a conceptual framework around which experiments could be conducted. Suggested experiments are provided at https://hackettmolecularbiology.blogspot.com/.

Students are encouraged to complete class activities, textbook readings and question sets in the order listed in each unit. Class activities, which are designed for use as class group work, are often intended to be completed before starting the related textbook readings so that students can discover concepts for themselves and thereby achieve the deepest level of understanding. These activities often assume that students will struggle to develop their answers without skipping ahead to future questions that may give away previous answers. It is the struggle itself that is key to effective learning. Instructors will ideally provide periodic feedback during these activities to guide students in their effective use of logic.

Table of Contents

ABOUT THE AUTHOR	iv
ACKNOWLEDGEMENTS	iv
PREFACE	v
HOW TO USE THIS INQUIRY WORKBOOK	v
TABLE OF CONTENTS	vi

UNIT 1

INTRODUCTION TO BIOCHEMISTRY AND CELL BIOLOGY	1
Class Activity 1A: Organic Molecules	2
Chapter 1 Reading #1 and Discussion Questions	7
Class Activity 1B: Monomers to Polymers	8
Chapter 1 Reading #2 and Discussion Questions	11
Class Activity 1C: Thermodynamics	12
Class Activity 1D: Nonspontaneous reactions in cells	17
Chapter 1 Reading #3 and Discussion Questions	20
Chapter 1 Reading #4 and Discussion Questions	21
Chapter 1 Reading #5 and Discussion Questions	22
Unit 1 Self-Assessment	23

UNIT 2

PROTEIN STRUCTURE AND FUNCTION	28
Chapter 2 Reading #1 and Discussion Questions	29
Class Activity 2A: The Role of Proteins in Genetic Disorders	30
Chapter 2 Reading #2 and Discussion Questions	34
Chapter 2 Reading #3 and Discussion Questions	34
Class Activity 2B: Fluorescent Protein Analysis	36
Chapter 2 Reading #4 and Discussion Questions	40
Chapter 2 Reading #5 and Discussion Questions	41
Chapter 2 Reading #6 and Discussion Questions	42
Unit 2 Self-Assessment	43

UNIT 3

DNA REPLICATION, REPAIR AND GENETIC ENGINEERING	47
Chapter 3 Reading #1 and Discussion Questions	48
Chapter 3 Reading #2 and Discussion Questions	49
Class Activity 3A: Inquiry into PCR	50
Chapter 3 Reading #3 and Discussion Questions	54
Chapter 3 Reading #4 and Discussion Questions	55
Chapter 3 Reading #5 and Discussion Questions	56
Class Activity 3B: Restriction Enzymes and Cloning	58
Class Activity 3C: Gibson Assembly Mutagenesis of Fluorescent Protein Genes	62
Unit 3 Self-Assessment	69

UNIT 4

THE REGULATION OF GENE EXPRESSION	75
Chapter 4 Reading #1 and Discussion Questions	76
Project 4A: Gene Parts - Analysis of an iGEM project	77
Class Activity 4B: Golden Gate Assembly of Yeast Gene Parts	80
Chapter 4 Reading #2 and Discussion Questions	85
Chapter 4 Reading #3 and Discussion Questions	86
Chapter 4 Reading #4 and Discussion Questions	87
Chapter 4 Reading #5 and Discussion Questions	88
Class Activity 4C: Epigenetic Inheritance	89
Chapter 4 Reading #6 and Discussion Questions	91
Unit 4 Self-Assessment	92

UNIT 5

GENOME EVOLUTION	95
Chapter 5 Reading #1 and Discussion Questions	96
Class Activity 5A: Cancer Genetics	98
Chapter 5 Reading #2 and Discussion Questions	104
Chapter 5 Reading #3 and Discussion Questions	105
Chapter 5 Reading #4 and Discussion Questions	106
Class Activity 5B: Telomeres and Mutation Rate	107
Chapter 5 Reading #5 and Discussion Questions	119
Unit 5 Self-Assessment	120

UNIT 6

EMERGING MOLECULAR BIOLOGY, BIOTECHNOLOGY AND MEDICINE	123
Chapter 6 Reading #1 and Discussion Questions	124
Class Activity 6A: Next Generation DNA Sequencing	125
Chapter 6 Reading #2 and Discussion Questions	126
Chapter 6 Reading #3 and Discussion Questions	127
Class Activity 6B: Curing a Genetic Disease with CRISPR/Cas9	128
Unit 6 Self-Assessment	131

PDB REFERENCES	132
APPENDIX 1: ANSWERS TO SELF-ASSESSMENTS	133
APPENDIX 2: BASIC MOLECULAR BIOLOGY TECHNIQUES	137
APPENDIX 3: *E. COLI* AND PHAGE PROMOTORS	144
APPENDIX 4: RESTRICTION ENZYMES	151
APPENDIX 5: REFERENCE INFORMATION	154

Unit 1

Introduction to Biochemistry and Cell Biology

We will begin by studying the intersection of biology and chemistry as it applies to biochemistry and cell biology. Hopefully you will recognize the majority of the basic concepts explored in this unit from your previous biology and chemistry courses, but they will be combined in a way that you may not have explored before – your goal should be to try to put all of the concepts together.

The teaching philosophy underlying inquiry learning is described in the "How to Use this Inquiry Workbook" section at the beginning of this book. To achieve the deepest level of understanding, you are encouraged to complete the class activities, textbook readings and question sets in the order listed below. **Class activities**, which are designed for use as class group work, are often intended to be completed before starting the related textbook readings so that you can discover concepts for yourself. You should not consult the textbook or other references when completing class activities, and instead should make use of your own logic during group discussions. You should thoroughly analyze individual questions in the order that they are given without skipping ahead. I advise covering the next page so you don't inadvertently short-circuit the learning process. Some of the questions are intentionally difficult and may require some struggle for your group. Your thoughtful, active engagement with the challenge will help you to achieve the deepest level of learning. Instructors will ideally provide periodic feedback to help guide your group during class activities. The **textbook readings and discussion questions** are intended to be assigned as homework, with completed question sets serving as starting points for class discussions. **Online quizzes** have also been developed for the purpose of self-checking reading comprehension of textbook readings. Check with your instructor about how to access their modified versions of the online quizzes for your course. Alternatively, if you are studying on your own, you may access online quizzes through the author's Google Classroom via the course code provided at https://hackettmolecularbiology.blogspot.com/. Suggested **experiments** that are posted at https://hackettmolecularbiology.blogspot.com/ are also listed here in brackets.

Suggested order:
    Class Activity 1A: Organic Molecules
    Chapter 1 Reading #1 and Discussion Questions
    [Experiment: Unknown Organic Molecules]
    Class Activity 1B: Monomers to Polymers
    Chapter 1 Reading #2 and Discussion Questions
    Class Activity 1C: Thermodynamics
    Class Activity 1D: Nonspontaneous Reactions in Cells
    Chapter 1 Reading #3 and Discussion Questions
    Chapter 1 Reading #4 and Discussion Questions
    Chapter 1 Reading #5 and Discussion Questions
    [Experiment: Biofuels/Alternative Energy]
    Unit 1 Self-Assessment Questions

Class Activity 1A: Organic Molecules

Work with others in your small class group to complete these questions. Thoroughly discuss one question at a time without skipping ahead. Use your logic skills to answer the questions - do not look up answers in your textbook or in another reference.

1. Obtain a Lewis Dot Kit consisting of cards to represent atoms (can be downloaded from https://hackettmolecularbiology.blogspot.com/) and bingo chips to represent valence electrons (or use pennies or other small items). Obtain a Periodic Table (in the Reference appendix of this book).

2. Recall from a previous Chemistry course:
   (Be sure to start by discussing this in your group - thinking about it is much better for your learning than just looking up the answers).
   a. What are valence electrons?
   b. How does one count the number of valence electrons in an atom?
   c. What are the rules for arranging valence electrons in Lewis Dot structures?

3. Using the cards and bingo chips in your kit, construct a Lewis Dot diagram showing the valence electrons in each molecule. After arranging the valence electrons in a molecule, **check with your instructor**, then copy the structure below.

   CH_4 	CH_3CH_2OH

   CH_3COOH 	CH_3NH_2
   Hint: O has the same number of lone pairs as above.

4. State the rules of bonding that you observe in the organic molecules above by completing the chart below:

Atom	# of bonds it makes	# of e- lone pairs	# of valence e- when bonded
hydrogen			
oxygen			
nitrogen			
carbon			

5. **Stop**. Check your answer to #4 with your instructor after discussing with your group.

6. Describe the exceptions to the rules above (for P or S) that should be remembered when studying organic molecules after a class discussion.

7. The drawings below represent the 4 main classes of organic molecules, except the bonds are **incorrect**. Here, every bond is drawn as a single bond. Sketch in the positions of any double or triple bonds that must be present.

8. Name the two atoms that are present in <u>all</u> organic molecules. _____

9. These organic molecules are drawn in a standard format in which not all atoms are shown. Propose how these drawings should be interpreted:
   a. Write in the missing atoms on structure I.
   b. Explain, in words, the rules for interpreting structures drawn like these.

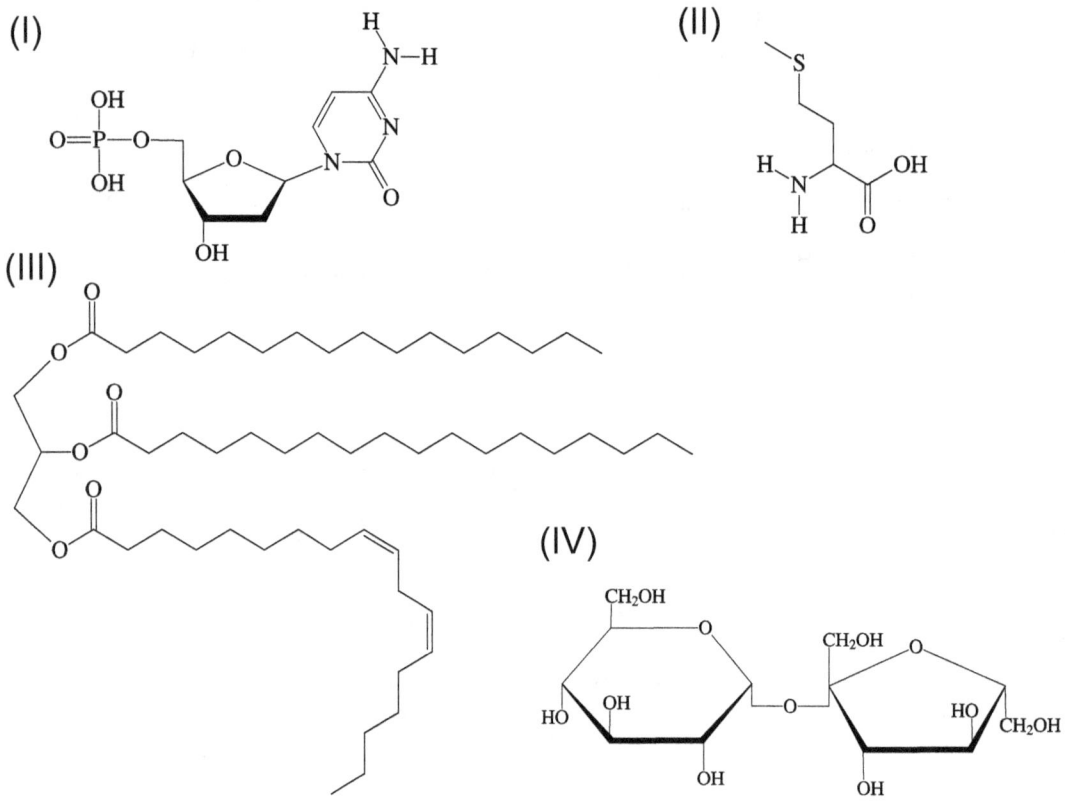

   c. How many hydrogens are in structure II? _____

10. Given the overall structures of these organic molecules, predict the solubility of each in water and predict whether the substance is hydrophobic or hydrophilic. Provide a chemical explanation for each of your answers (using your period table and the electronegativity chart in the reference section).

   I.

   II.

   III.

   IV.

11. What principles from Chemistry did you use to arrive at your decisions in the previous question?

12. What happens on a molecular level when a substance dissolves in water?

13. Written below are some examples of **inorganic** acids and bases:

    HCl + H₂O → H₃O⁺ + Cl⁻

    NH₃ + H₂O → NH₄⁺ + OH⁻

    What do you remember from Chemistry is the definition of a Brønsted-Lowry acid? A base?

    a. Acid:

    b. Base:

14. **Organic** molecules sometimes include special functional groups (segments with specific chemical behavior) that act as acids or bases. Below, the squiggles in the structures represent the position where the functional group would connect to the rest of an organic molecule. Identify the functional group(s) below that act as acids. Identify the functional group(s) that act as bases. How do you know?

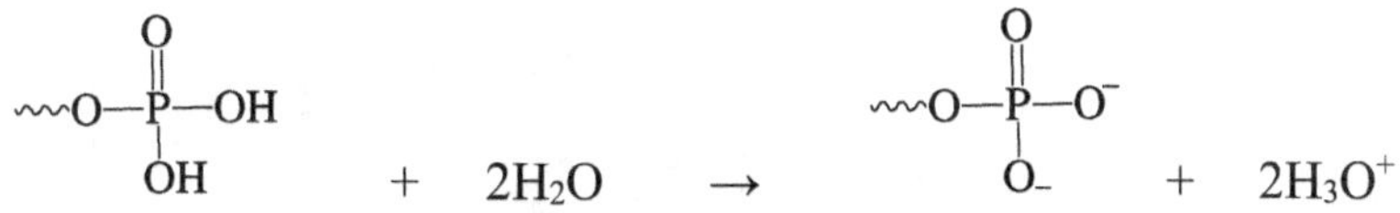

15. Choose one of the functional groups above and use your Lewis Dot Kit to model the reaction with water. What is transferred?

16. The functional groups above are the phosphate group, carboxylic acid group, and amino group. State your best guess for the identity of each – label them above.

17. In a cell, would you expect acidic and basic functional groups to be charged or uncharged? Why?

18. For the structures drawn below, (a) circle and label any acidic or basic functional groups. (b) Box and label any large hydrophobic domains. (c) Challenge: Identify the carbohydrate, lipid, fatty acid, amino acid, protein, and nucleic acid, based on your interpretation of these names and based on your previous understanding of the chemistry of these molecules (i.e. soluble in water?).

(I)

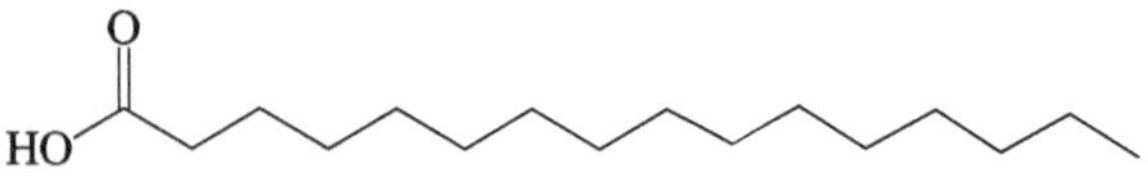

(II)

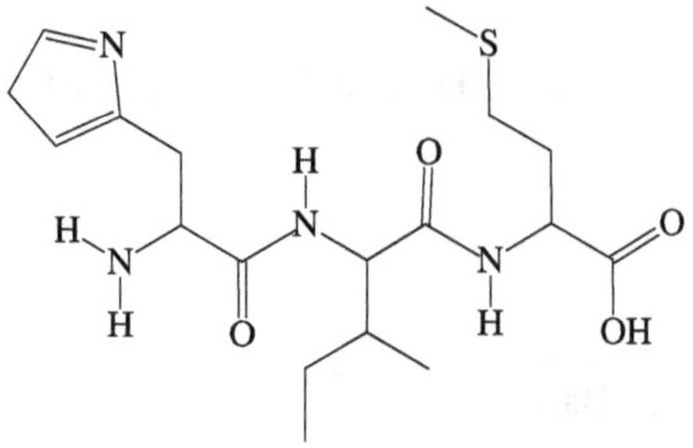

(III)

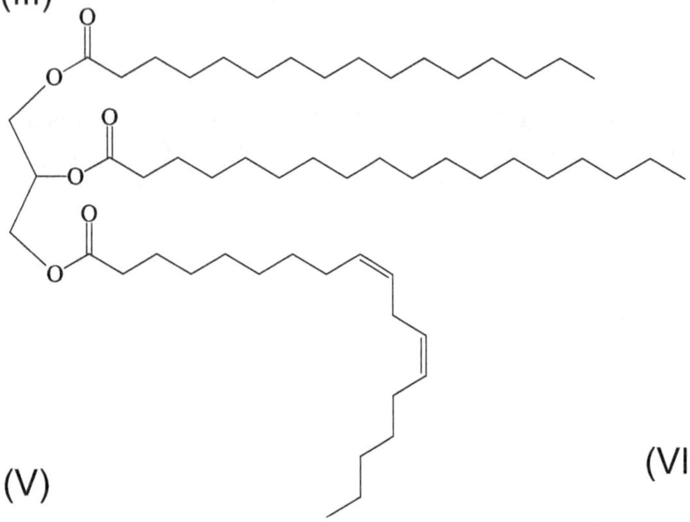

(IV)

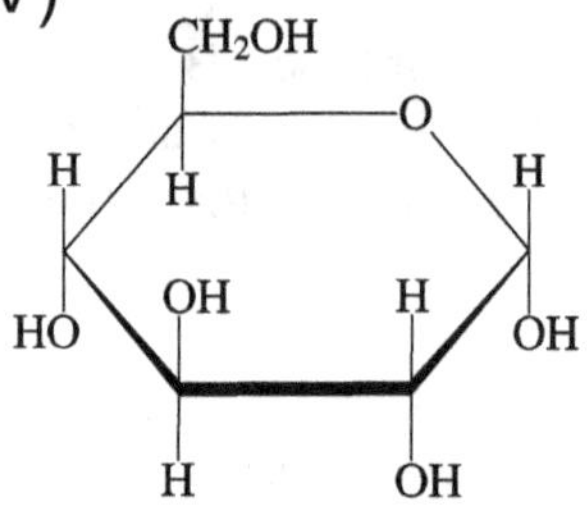

(V)

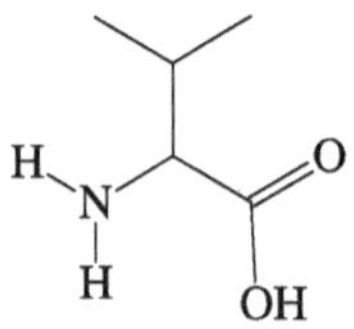

(VI)

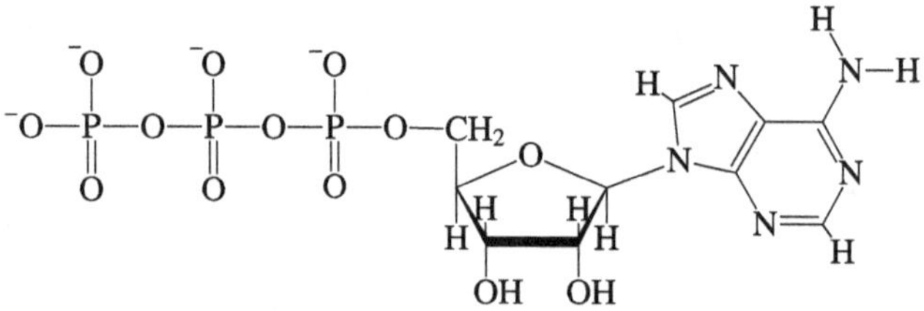

Class Activity 1A: Organic Molecules

Chapter 1 Reading #1 and Discussion Questions

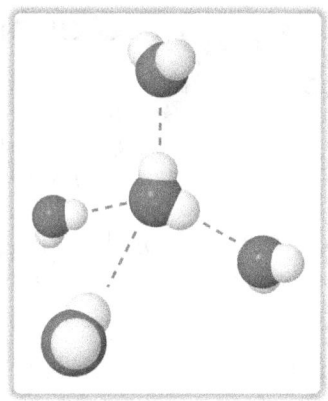

Reading: *Molecular Biology: Concepts for Inquiry*
Chapter 1: Introduction and Sections 1.1A – 1.1D

Online Quiz: 1.1A – 1.1D
The online quizzes are intended to help you self-assess your fundamental comprehension of the reading. Complete the online quiz before answering the discussion questions below. Review relevant textbook sections if you miss questions. Check with your instructor about how to access their modified versions of the online quizzes for your course. Alternatively, if you are studying on your own, you may access online quizzes through the author's Google Classroom via the course code provided at https://hackettmolecularbiology.blogspot.com/.

Discussion Questions:
The class discussion questions allow you to analyze the textbook readings more deeply. Some questions directly assess your understanding of the reading, 📖, while others ask you to apply what you have learned, ⚙. A few questions are intended to help you think about your thinking, 🦉, because metacognition can deepen your understanding.

1. 📖 What do you think is the role of engineering in biotechnology?

2. ⚙ Why do you think living things have evolved to use carbon (and not another atom) as the backbone of organic molecules?

3. ⚙ Draw one organic molecule that's not in your textbook. Also draw one inorganic molecule. How can you tell by looking at the structures which is organic and which is inorganic?

4. 🦉 Explain how an understanding of electronegativity can enhance one's understanding of biological systems.

5. ⚙ Why can water dissolve some substances, but not others?

6. 🦉 In Chemistry, you likely studied the role of inter-particle electrostatic forces in holding together a solid or liquid pure substance (in terms of their effect on boiling point, etc). However, biological systems consist of complex aqueous solutions of thousands of kinds of molecules, and in some places even heterogeneous mixtures. What did you learn about these electrostatic forces in Chemistry that is relevant to an understanding of electrostatic forces in biological systems? What new principles are you now integrating into this understanding?

Class Activity 1B: Monomers to Polymers

Work with others in your small class group to complete these questions. Thoroughly discuss one question at a time without skipping ahead. Use your logic skills to answer the questions - do not look up answers in your textbook or in another reference.

1. Partial chemical reactions involving monomers and polymers are shown below. Determine what's missing in each reaction and add it below.

(I)

(II)

(III)

8 | Unit 1   *Class Activity 1B: Monomers to Polymers*

(IV)

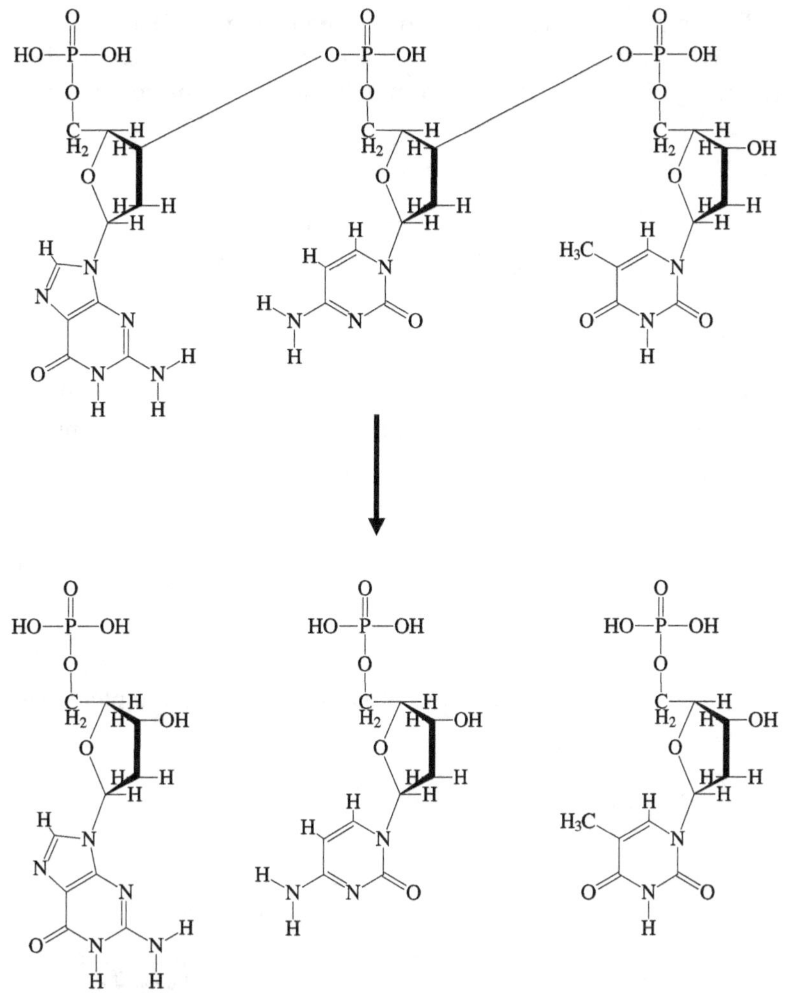

2. a. Which of the following cartoons correctly shows dehydration synthesis?

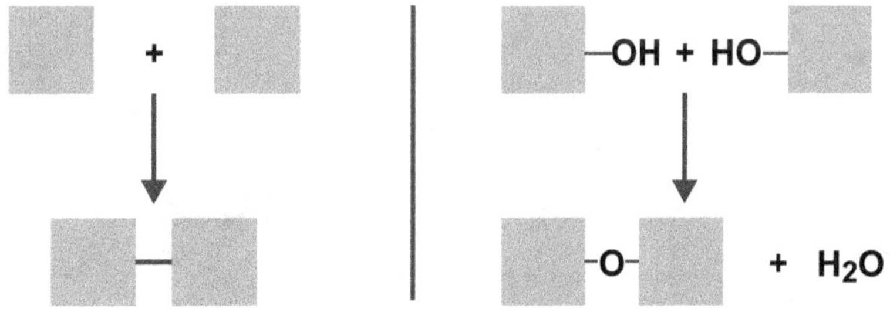

b. Construct your own generalized diagrams (using shapes to represent molecules and showing only relevant functional groups/atoms) to describe the general processes by which polymers form monomers.

c. Determine the names of these two general processes: Discuss in your group and if no one remembers from a Biology course, look it up in section 1.1E of your textbook. Explain why these names make sense.

3. Hypothesize: One of these processes tends to produce free energy (ΔG) and the other process tends to require an input of free energy. Hypothesize about which is which, given your general understanding of organic molecules and/or your understanding of thermodynamics from Chemistry. If you're not sure, don't worry...we're studying this next.

4. Besides the reactants and sometimes an input of free energy, what else do you think is always present when these reactions occur in your body? Explain. Why is it a good thing that these reactions are extremely slow otherwise?

5. Now apply what you've learned to complete this partial reaction. The complete reaction is the hydrolysis of ATP. Name each molecule.

Chapter 1 Reading #2 and Discussion Questions

Reading: *Molecular Biology: Concepts for Inquiry*
Chapter 1: Sections 1.1E – 1.1F

Online Quiz: 1.1E – 1.1F

Discussion Questions:

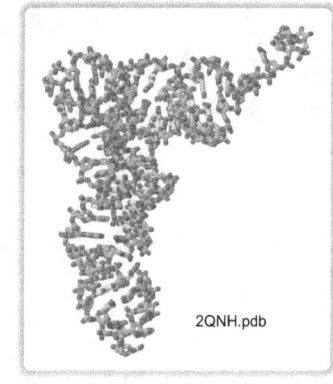

1. Why are most carbohydrates soluble in water? Why isn't cellulose? Provide a real-world example of how humans have taken advantage of the fact that cellulose is insoluble.

2. Where do you think hydrolysis takes place in the human body? Dehydration synthesis? Explain your reasoning.

3. Cells don't have any special machinery for constructing a cell membrane from phospholipids. Explain how you think the cell membrane forms.

4. In terms of their structures, explain why proteins are considered to be the most diverse class of biological molecule.

5. In Figure 1.12, other than mRNA, the other types of RNA molecules contain double-stranded regions. Hypothesize about why these RNA molecules might have evolved to have double-stranded regions.

6. Why are DNA-binding regions of proteins frequently positively charged?

7. Given the examples in this chapter, which kinds of functional groups are able to participate in dehydration synthesis reactions? What do they have in common?

Class Activity 1C: Thermodynamics

Work with others in your small class group to complete these questions. Thoroughly discuss one question at a time without skipping ahead. Use your logic skills to answer the questions - do not look up answers in your textbook or in another reference.

1. Complete the following table:

	Animal	Plant
In what form(s) does an organism take in energy?		
In what form(s) does an organism store energy?		
In what form(s) does an organism give off energy?		

2. Your thoughts now: When we say that a molecule has chemical potential energy, what does that mean? Why is a molecule like glucose considered to be an energy source to fuel an organism, but a molecule like carbon dioxide is not? Or, what must be true for the chemical potential energy stored in a molecule to be made available to a cell? (You'll consider this question again at the end of this activity - do your best here to capture your current understanding).

The change in enthalpy for a reaction can be determined from a potential energy diagram:
ΔH = H$_{products}$ - H$_{reactants}$

a) **Exothermic Reaction (ΔH is negative)**   b) **Endothermic Reaction (ΔH is positive)**

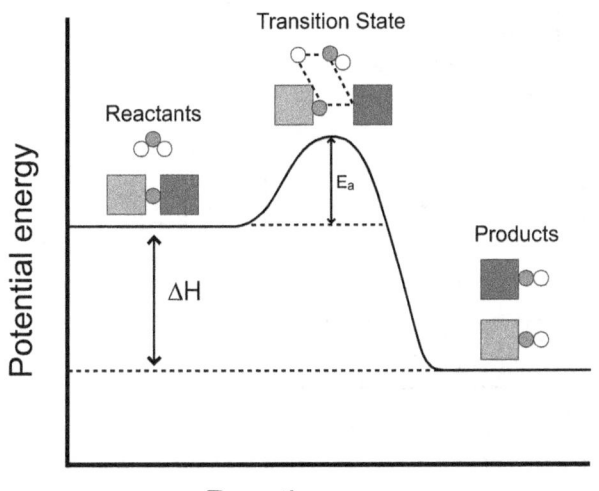 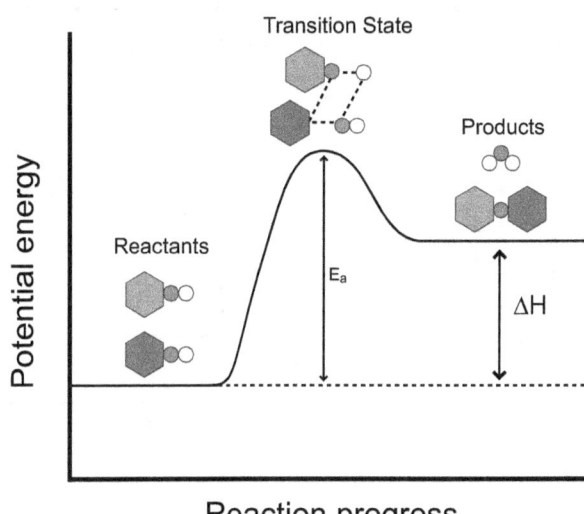

3. Based on the diagrams above, complete the following statements:

    Breaking bonds_____ energy.

    Making bonds_____ energy.

4. The enthalpy of a molecule is related to its chemical potential energy. The enthalpy is determined by the number, type, and arrangement of bonds within a molecule. For example, a C-C single bond requires less energy to break and releases less energy when it forms than a C=C double bond. The more stable the arrangement of atoms in a molecule, the lower its chemical potential energy.

    Would you predict that the change in enthalpy, ΔH, would be positive or negative for a reaction in which stronger/more bonds are present in the products than in the reactants? Explain.

Understanding enthalpy is only part of the answer to #2. You must also consider changes in entropy.

5. Discuss in your group: What do you remember about entropy from previous chemistry courses?

6. For each of the following pairs of drawings, circle the one that you think has the greater entropy. Can you explain why? [Drawings 1a and 1b provided by Julia Masters; folded protein 3MW6.pdb].

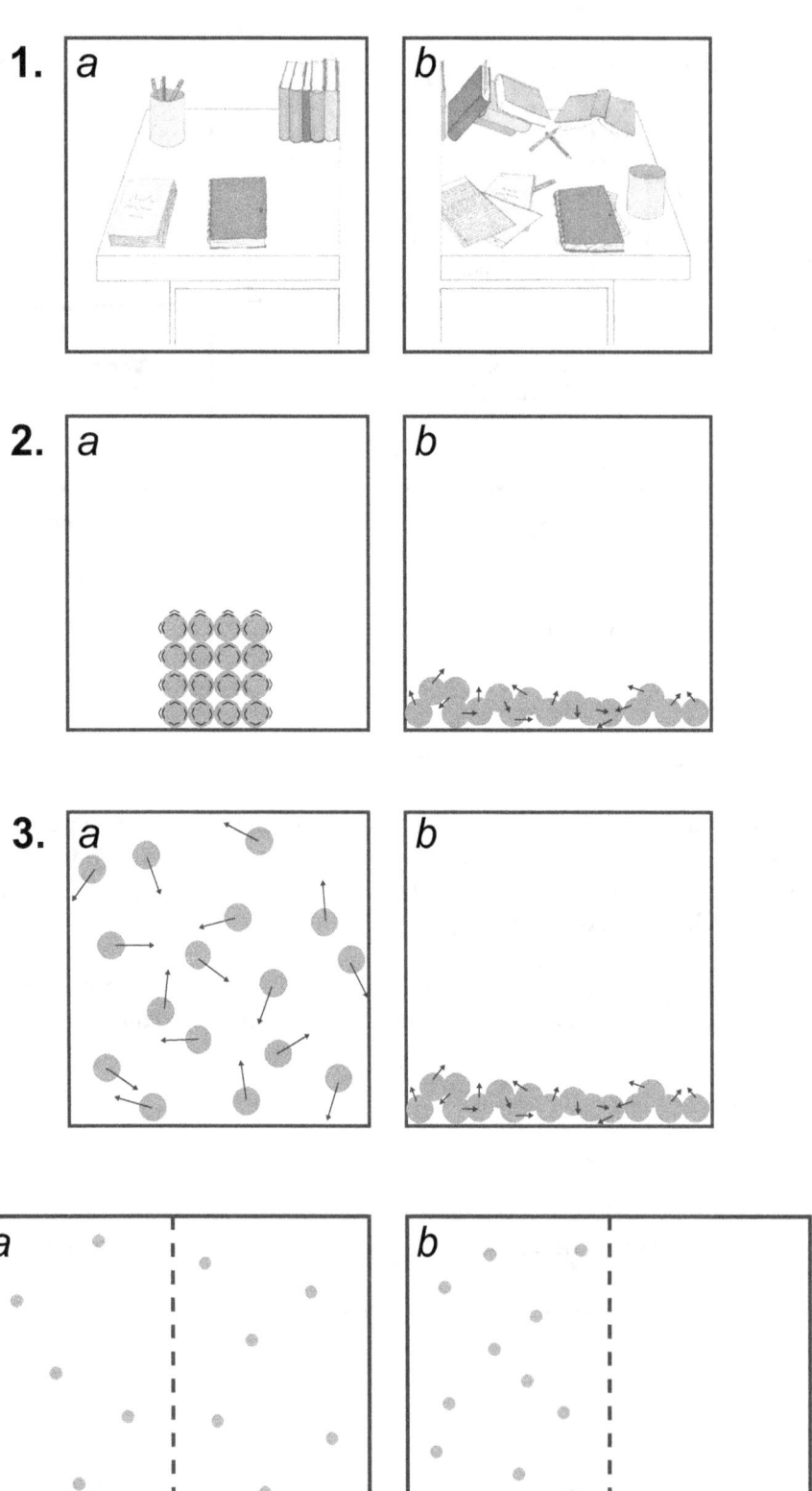

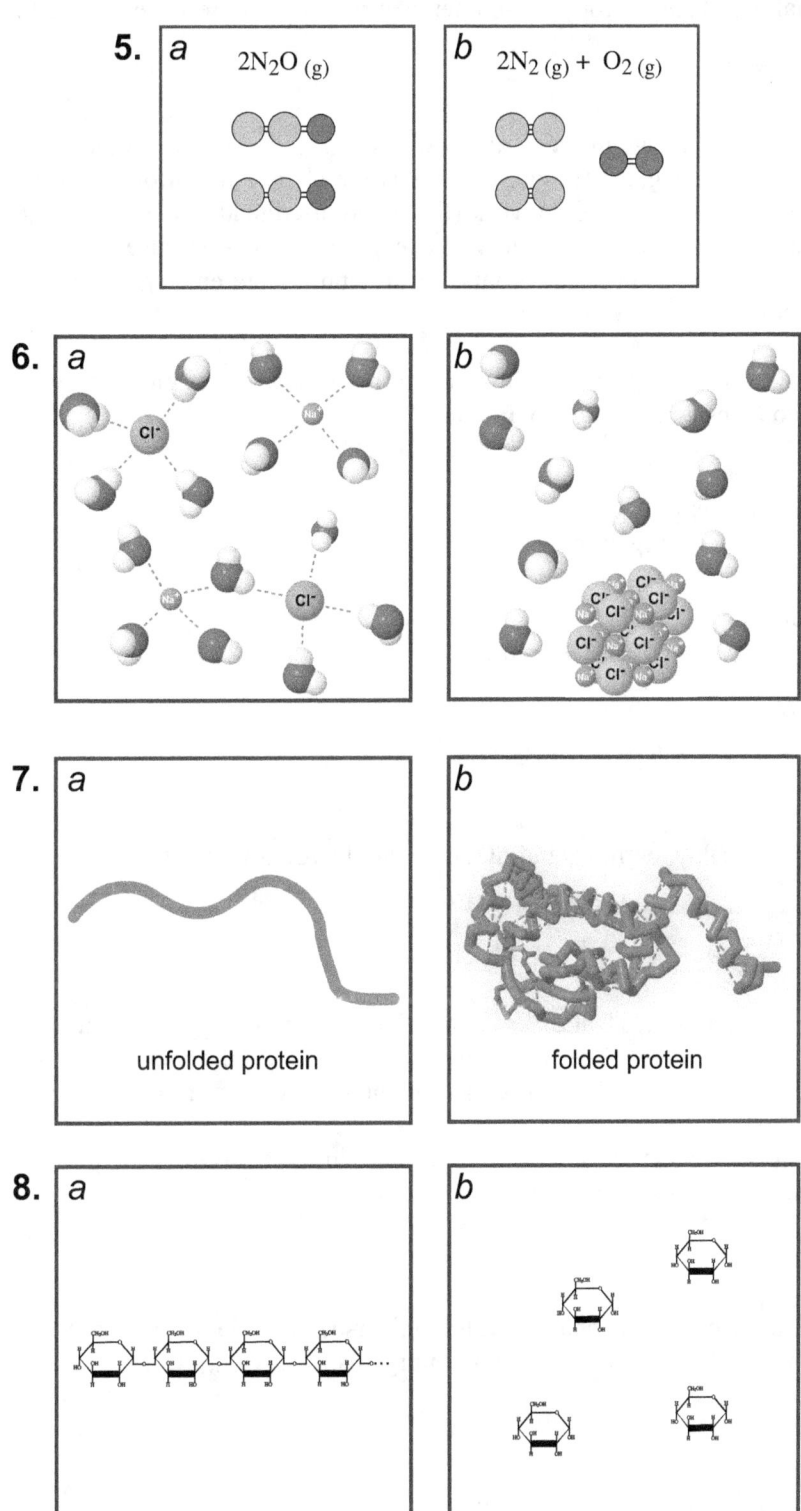

7. After a class discussion, how would you define entropy?

Class Activity 1C: Thermodynamics Unit 1 | 15

The contributions of enthalpy (H) and entropy (S) to chemical reactions are reflected in the following formula:

$$\Delta G = \Delta H - T\Delta S$$

In the above equation, ΔG is the change in **free energy** of a chemical reaction or other process. ΔH is the change in enthalpy, T is the temperature in degrees Kelvin, and ΔS is the change in entropy. ΔG, the change in free energy, quantifies the portion of ΔH that is "free" to do work (that is not used/made as entropy changes). If ΔG is a negative value, then the process occurs spontaneously and it is an **exergonic** process because it gives off free energy. If ΔG is a positive value, then the process does not occur spontaneously and it is an **endergonic** process because it requires an input of free energy to occur.

8. Think back to what you learned in Biology – for each of the following reactions, do you think that ΔG is positive or negative? (Check your answers before moving on).

ΔG

Hydrolysis of ATP:
ATP + H_2O → ADP + P_i

Photosynthesis:
$6 CO_2 + 6 H_2O \rightarrow C_6H_{12}O_6 + 6 O_2$

Cellular Respiration:
$C_6H_{12}O_6 + 6 O_2 \rightarrow 6 CO_2 + 6 H_2O$

9. Now, having reviewed enthalpy, entropy, and ΔG, how would you revise your answer to #2?

The First Law of Thermodynamics states that $\Delta E_{universe}$, the change in the energy of the universe, is always equal to zero. This means that if the energy of a system, like an organism, increases, then the energy of its surroundings must decrease proportionally. (If you eat food, you take in energy, and it came from your surroundings).

10. The Second Law of Thermodynamics describes the change in the entropy of the universe. Use your reasoning to complete the following equation with <, >, or =.

$\Delta S_{universe}$ _____ 0

11. How does a living thing decrease the entropy of the universe? How does it increase the entropy of the universe? Since life exists, what must be the overall effect of an organism on $\Delta S_{universe}$?

Class Activity 1D: Nonspontaneous reactions in cells

Work with others in your small class group to complete these questions. Thoroughly discuss one question at a time without skipping ahead. Use your logic skills to answer the questions - do not look up answers in your textbook or in another reference.

1. List real-world examples of spontaneous processes.

2. List real-world examples of nonspontaneous processes.

3. Since we know some nonspontaneous processes can happen, what do you think is necessary to make them happen?

4. In general, reactions that build polymers from monomers are nonspontaneous. One example of such a nonspontaneous reaction is the dehydration synthesis of maltose from glucose:

    $$2 \text{ glucose} \rightarrow \text{maltose} + H_2O \qquad \Delta G° = +15 \text{ kJ/mol}$$

    Since this process happens in cells, what do you think might happen to drive this nonspontaneous process?

The nonspontaneous dehydration synthesis of maltose:
$$2 \text{ glucose} \rightarrow \text{maltose} + H_2O \qquad \Delta G° = +15 \text{ kJ/mol}$$
is **coupled** to the spontaneous hydrolysis of 2 ATP:
$$2 \text{ ATP} + H_2O \rightarrow 2 \text{ ADP} + PP_i \qquad \Delta G° = -29 \text{ kJ/mol}$$

The synthesis of maltose from glucose: Coupled reactions showing intermediates and enzymes

#1 Glucose + ATP + H₂O →(Hexokinase) Glucose-6-Phosphate + H⁺ + ADP + H₂O

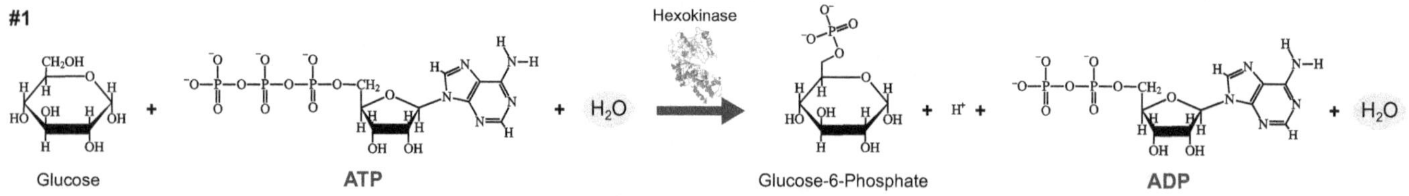

#2 Glucose-6-Phosphate + H₂O →(Phosphoglucomutase) Glucose-1-Phosphate + H₂O

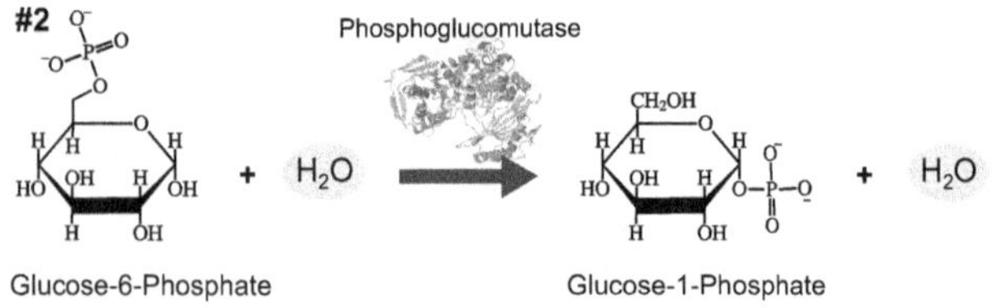

#3 Glucose-1-Phosphate + ATP + 2H₂O →(ADP-glucose pyrophosphorylase) ADP-Glucose + PP$_i$ + 2H₂O

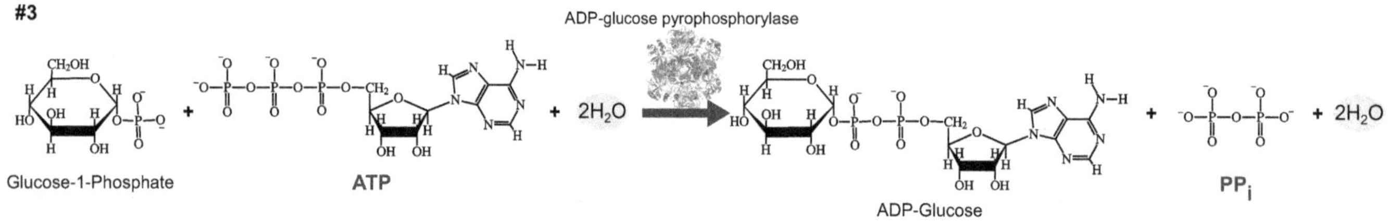

#4 ADP-Glucose + Glucose + H₂O →(Glycogen phosphorylase) Maltose + H⁺ + ADP + H₂O

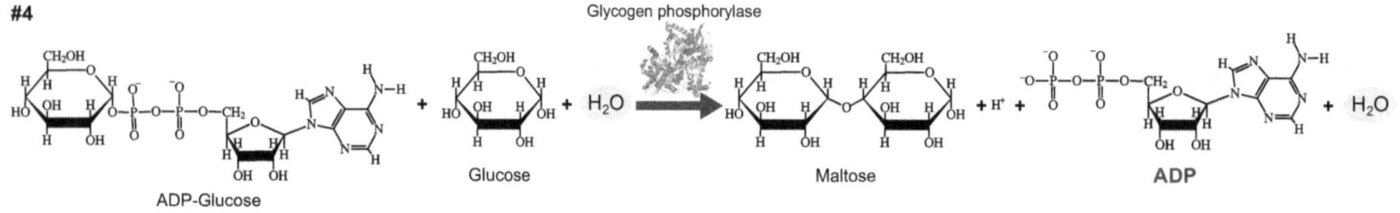

5. In the above example, circle the initial reactants and final products. Do not circle any intermediate compounds that are made, but later used (it may help to cross these out). Write the overall, simplified, coupled reaction for the synthesis of maltose, ignoring any intermediate compounds that are made, but later used. Also solve for the ΔG of the overall process (the overall ΔG is the sum of the individual ΔGs, regardless of the identities of any intermediate compounds). Enzyme structures above are from PDB files 1HKG, 3PMG, 1YP2, and 8GPB.

6. Is the overall coupled reaction spontaneous? How do you know?

7. The above example is typical of how a nonspontaneous reaction is coupled to a spontaneous reaction in cells. What characteristics of coupled reactions do you notice?

8. Your instructor will provide several cards depicting nonspontaneous and spontaneous processes (or you can obtain them from https://hackettmolecularbiology.blogspot.com/). Match up the cards in a way that would allow all of the nonspontaneous processes to take place. Use your logic skills and prior knowledge of biology to do this.

   Check with your teacher that your logic makes sense, then obtain the second set of cards with the names of the processes and match these to the coupled reactions.

   You can record your answers in the table below:

Nonspontaneous process		Spontaneous process		Coupled	
Process	ΔG (kJ/mol)	Process	ΔG (kJ/mol)	Process	ΔG (kJ/mol)
2 glucose $\rightarrow$ maltose + H_2O	+15	2 ATP + H_2O $\rightarrow$ 2 ADP + PP_i	-29	Dehydration synthesis	-14
DNA (4 nt) + nucleoside monophosphate $\rightarrow$ DNA (5 nt) + H_2O	+22				
Polypeptide (5 amino acids) + amino acid $\rightarrow$ Polypeptide (6 amino acids) + H_2O	+21				
Transport of 3 Na^+ ions out of a cell and 2 K^+ ions into a cell against their concentration gradients using a Na^+/K^+ protein pump.	+37				
38 ADP + 38 P_i $\rightarrow$ 38 ATP	+1159				
6 CO_2 + 6 H_2O $\rightarrow$ glucose + 6 O_2	+2867				

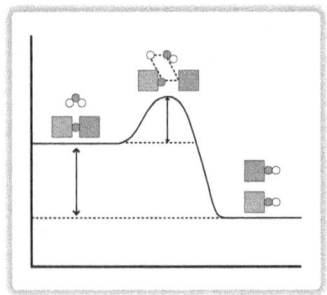

Chapter 1 Reading #3 and Discussion Questions

Reading: *Molecular Biology: Concepts for Inquiry*
Chapter 1: Section 1.2

Online Quiz: 1.2

Discussion Questions:

1. Why are ΔE_{system} and $\Delta E_{surroundings}$ always of equal magnitude, but opposite sign?

2. How do organisms gain energy? Is it different for different types of organisms? How do organisms lose energy?

3. Give a real-world example of a system in which entropy is increasing. Provide an explanation as to why entropy is considered to have increased in your example.

4. Explain how life is compatible with the second law of thermodynamics.

5. Explain the difference between free energy and enthalpy.

6. In terms of your understanding of energy and thermodynamics, why does it make sense that a process where ΔH is negative and ΔS is positive would always occur spontaneously? (Don't just restate the equation – think about what it means).

7. In Table 1.5 (Examples of spontaneous cellular processes), all of the processes shown are ones in which entropy decreases. What makes these processes spontaneous?

8. Explain how cells are able to carry out nonspontaneous reactions.

9. Given reaction A (ΔG = 400 kJ/mol) and reaction B (ΔG = -500 kJ/mol), identify the spontaneous and nonspontaneous reactions. If these reactions are coupled, will the overall reaction occur? What is the ΔG for the combined process? What features might you expect to observe for the coupled reactions?

10. ⚙ What is the role of enzymes in chemical reactions? What are possible misconceptions about the roles of enzymes that you should avoid?

Chapter 1 Reading #4 and Discussion Questions

Reading: *Molecular Biology: Concepts for Inquiry*
Chapter 1: Section 1.3

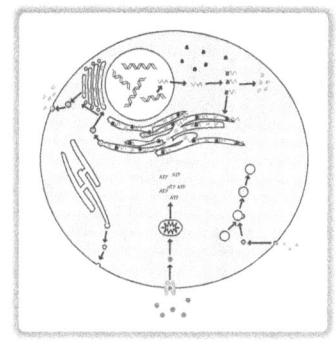

Online Quiz: 1.3

Discussion Questions:

1. 🔲 Explain how the coupling of spontaneous and nonspontaneous reactions allows an organism to meet its energy needs.

2. 🔲 What does it mean for a chemical reaction to be in a state of dynamic equilibrium? What misconceptions about equilibrium should you avoid?

3. ⚙ Explain the function of organelles in terms of the thermodynamics of a cell. Why don't bacteria need organelles?

4. ⚙ Give an example of how multiple organelles or other cellular structures might function together to carry out a complex process.

5. ⚙ Explain the thermodynamics of active transport.

6. ⚙ Consider the functions of the individual organelles. How does each organelle contribute to the thermodynamic efficiency of the cell?

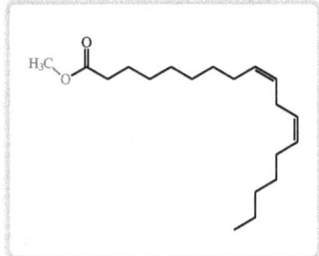

Chapter 1 Reading #5 and Discussion Questions

Reading: *Molecular Biology: Concepts for Inquiry*
Chapter 1: Section 1.4

Online Quiz: 1.4

Discussion Questions:

1. Why is the development of alternative energy sources to power jets more challenging than producing energy to power homes or cars?

2. Explain how nonspontaneous and spontaneous processes are coupled to allow the production and use of ethanol as an alternative energy source.

3. Why is ethanol more environmentally friendly than gasoline?

4. Explain the drawbacks to using ethanol as a fuel source.

5. What are the advantages and disadvantages of the use of cellulose as the starting material for ethanol production compared to the use of glucose or starch?

6. Explain the role of electrostatic forces in the methods used to purify biodiesel.

7. Why is the miscibility of ethanol and water problematic in terms of the use of ethanol as a fuel? How is genetic engineering being used to address this problem?

8. How are the processes that occur in the "artificial leaf" similar to and different from the processes that occur in photosynthesis?

9. If you were asked to advise the government about how to most effectively allocate funding for research into the development of alternative fuel sources, what would you advise and why?

Unit 1 Self-Assessment

1. Given the following organic molecule:

   a. Circle any polar bonds in this molecule.

   b. Label any atoms with a partial negative charge δ⁻ and any atoms with a partial positive charge δ⁺.

   c. How many valence electrons does the double-bonded oxygen atom control (partially or fully)?

   d. Explain why the functional group in this molecule is described by biochemists as behaving as either an acid or a base (state what it does that is acid-like or base-like).

2. Part of a protein called the T-cell receptor is embedded in the cell membrane. What would you predict would be true about the polarity of the portions of the protein that face into the middle of the membrane? Explain.

3. Given the following molecule:

   a. How many hydrogen atoms does this molecule contain?

   b. Name two functional groups present in this molecule.

   c. To which of the four classes of biological molecules does this molecule belong?

   d. Should this molecule dissolve in water?

4. Complete the following table:

Monomer(s)	Polymer(s)
	lipids
amino acid	
	glycogen, starch, cellulose
nucleotides	

5. Given that starch and cellulose are both polymers of glucose, explain why they differ in terms of whether they dissolve in water.

6. When water and oil are mixed, they separate into layers. Explain why in terms of polarity and intermolecular forces.

7. Based on the structure of DNA, which do you think has the stronger attractions between the nitrogenous bases: an A-T base pair or a G-C base pair?

8. Sketch the Lewis Dot structure for the phosphate ion, PO_4^{3-}.

9. The following reaction is incomplete. Add the missing reactants/products.

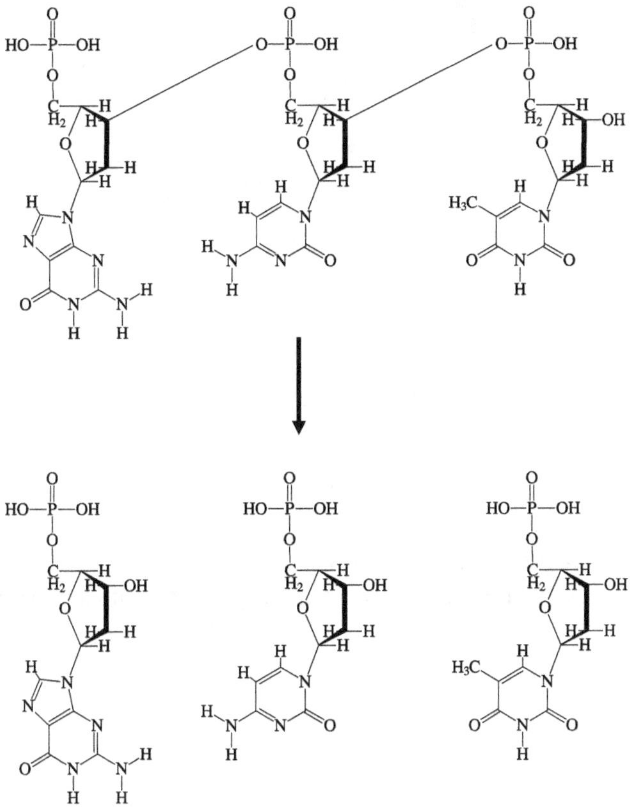

10. Sketch the product(s) that will result if these molecules undergo dehydration synthesis.

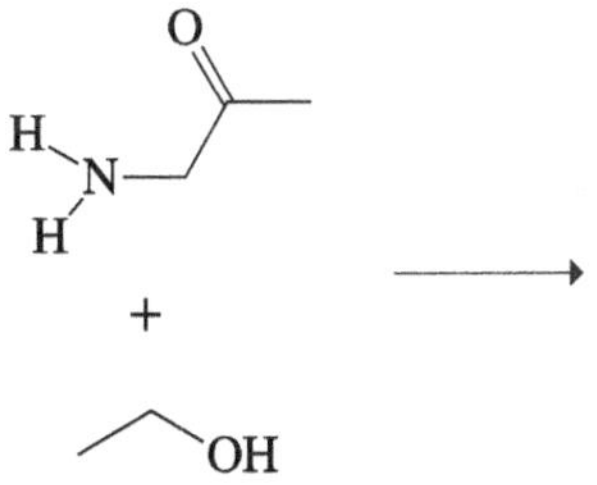

11. Complete the following statements to indicate which process releases energy and which process requires an input of energy:

Breaking bonds _____ energy.

Making bonds _____ energy.

12. a. Protein folding is a spontaneous process. Explain in terms of ΔH, ΔS and ΔG. For each, explain why the value changes in this direction.

    b. Under what condition would this process not be spontaneous?

13. If the reaction of "A" + "B" to form "C" has a standard ΔG° of +10kJ/mol, is it possible for this reaction to occur in a cell if it is not coupled to a spontaneous process? If not, explain why not. If yes, explain what would be necessary for it to occur.

14. a. State common features of coupled reactions in cells. Be sure to indicate the role of enzymes.

    b. What is the purpose of fermentation in yeast (the purpose for the yeast, not for humans)?

15. Explain in terms of thermodynamics why it makes sense that bacteria have not evolved organelles, but eukaryotic cells have.

16. Explain why humans perform the hydrolysis of some type(s) of glucose polymers, but not other type(s) of glucose polymers.

17. Each of the following cells contains extra copies of a particular organelle compared to the average cell. Name the organelle present in extra copies for each cell:
    a. This cell in the pancreas produces large amounts of the protein hormone, insulin, which is exported to the blood.
    b. This cell in the immune system engulfs bacteria and fuses them with this organelle to hydrolyze their organic molecules.
    c. The proteins in this muscle cell require ATP to change shape during muscle contraction.
    d. This kind of cell produces glucose for the rest of the organism.

18. True/False:
    a. Adding an enzyme will increase the spontaneity of a reaction.
    b. All chemical reactions that give off heat are spontaneous.
    c. The hydrolysis of ATP is endothermic since bonds are broken.
    d. If ΔG for the reaction A→ E is -300kJ/mol, the overall ΔG for the reaction series A→ B→ C→ D→ E is -300kJ/mol.
    e. The entropy of a starch polymer containing 100 glucose monomers is greater than the entropy of 100 individual glucose monomers.

19. Identify the organelle that best fits each description.

Organelle	Description
	Couples the input of light energy to the nonspontaneous synthesis of glucose.
	Couples the combustion of glucose to ATP synthesis
	In this organelle, the addition of nucleoside monophosphates to a nucleic acid chain is coupled to the hydrolysis of nucleoside triphosphates.
	Its main function is the hydrolysis of polymers into monomers
	Site of the synthesis of proteins for export.
	Site of phospholipid synthesis

20. In the combustion of glucose, which group of molecules has the lower chemical potential energy, reactants or products? Explain, in terms of bonding, why these molecules have the lower chemical potential energy.

Unit 2

Protein Structure and Function

Unit 1 introduced protein structure and some of the roles of proteins in cells. In this chapter, we will explore protein structure and function in more depth.

To achieve the deepest level of understanding, you are encouraged to complete the class activities, textbook readings and question sets in the order listed below.

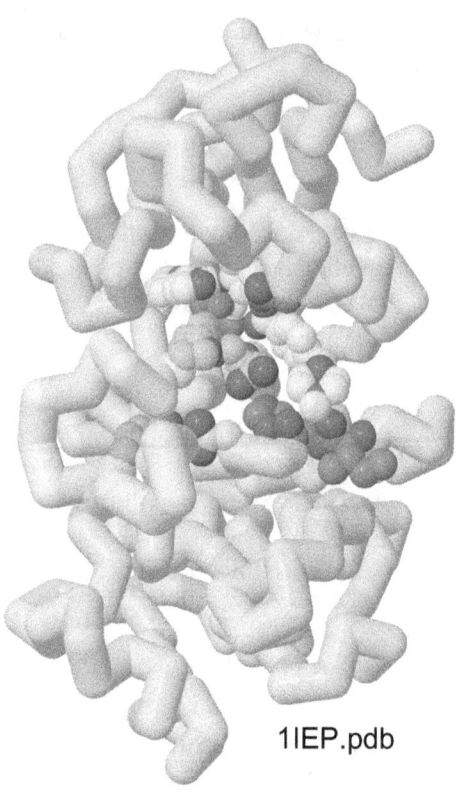

1IEP.pdb

Suggested order:
    Chapter 2 Reading #1 and Discussion Questions
    Class Activity 2A: The Role of Proteins in Genetic Disorders
    Chapter 2 Reading #2 and Discussion Questions
    Chapter 2 Reading #3 and Discussion Questions
    [Experiment: Fluorescent Protein Analysis]
    Class Activity 2B: Fluorescent Protein Analysis
    Chapter 2 Reading #4 and Discussion Questions
    Chapter 2 Reading #5 and Discussion Questions
    Chapter 2 Reading #6 and Discussion Questions
    Unit 2 Self-Assessment Questions

Prior knowledge (review if necessary):
- Structure of DNA, RNA, and proteins (in Chapter 1).
- Complementary DNA and RNA nucleotides (in Chapter 1)
- Coupling of spontaneous processes to nonspontaneous processes (in Chapter 1)
- Thermodynamic principles: free energy, enthalpy, entropy (in Chapter 1)

Chapter 2 Reading #1 and Discussion Questions

Reading: *Molecular Biology: Concepts for Inquiry*
Chapter 2: Introduction and Section 2.1A

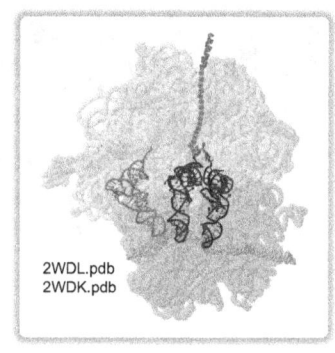

Online Quiz: 2.1A

Discussion Questions:

1. Given the following sense strand of DNA, determine the sequence of the antisense strand of DNA, the sequence of the mRNA and the sequence of the polypeptide that is encoded:
5'CCAATGGCAGAATGACCT3'

2. Why are the ends of the DNA strands marked 5' and 3'?

3. What does it mean if a gene is "expressed"?

4. Compare and contrast the steps in transcription, translation, and pre-mRNA processing in prokaryotes and eukaryotes. How do prokaryotic mRNAs differ from eukaryotic mRNAs?

5. How do introns contribute to the complexity of eukaryotes compared to prokaryotes?

6. Why do you think the genetic code evolved to be non-random?

7. Biosynthetic pathways are generally nonspontaneous. What reactions are coupled to those in transcription and translation to make these processes possible? Why is the ribosome needed for translation?

8. In what ways is RNA important to transcription and translation?

9. You decide to express the gene for human insulin in bacteria. You copy the gene from human DNA, insert it into a circle of bacterial DNA and put the DNA into the bacteria. Will the bacteria express insulin? Explain.

10. Explain why bacterial ribosomes are good drug targets for antibiotics.

Class Activity 2A: The Role of Proteins in Genetic Disorders

Work with others in your small class group to complete these questions. Thoroughly discuss one question at a time without skipping ahead. Use your logic skills to answer the questions - do not look up answers in your textbook or in another reference.

1. First, discuss the following review questions with your group. If your group is unfamiliar with any of the terms, consult your instructor before moving on.

    a. What is the difference between "DNA," "gene," and "chromosome"?

    b. What is an allele? How many alleles of each gene does a person have?

    c. What's the difference between a somatic cell and a gamete?

    d. How many chromosomes does a human have in each somatic cell?

    e. Where did these chromosomes come from?

    f. What is meant by a "pair" of chromosomes?

    g. What do "dominant" and "recessive" mean?

    h. What's the difference between a genotype and a phenotype?

    i. What's the difference between "homozygous" and "heterozygous"?

    j. What is a "wildtype" allele?

2. In the following pedigrees, squares represent males, circles represent females, open shapes indicate a normal phenotype, and filled shapes represent a disease phenotype. For each pedigree, indicate whether the **more likely** mode of inheritance is dominant or recessive.

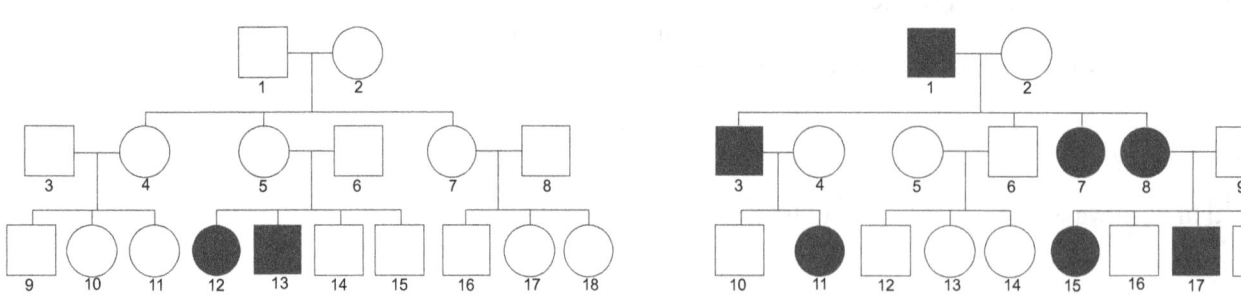

3. How do you decide which mode of inheritance is more likely if both are possible?

<u>STOP</u>. Check your response before moving on.

In the following exercises, we will explore the kinds of mutations in genes and types of changes in proteins that might result in either dominant or recessive disease phenotypes.

Mutations can be grouped into two broad categories:
a. Loss of function – The gene mutation reduces or eliminates the function of the protein. The disease results from the absence of one or two doses of the normal protein.
b. Gain of function – The gene mutation causes the protein to function abnormally. The disease results from the presence of one dose of a mutant version of the protein.

4. Which type(s) of mutations do you think might create a recessive phenotype? Explain.

5. Which type(s) of mutations do you think might create a dominant phenotype? Explain.

6. If you knew which gene had been mutated in a patient, but didn't know the nature of the mutation, which of the following do you think would be the best way to determine whether the mutation produced a loss of function or a gain of function of the protein? Why?
   a. Examine the patient's pedigree.
   b. Treat the patient with a drug.
   c. Culture the patient's cells. Reintroduce a normal (wildtype) copy of the gene into the cells. Study whether or not the normal phenotype is restored.
   d. Sequence the gene in the patient to identify the DNA mutation.
   e. More than one of the above.

7. For each description of a genetic disease below, identify whether the genetic disorder is most likely dominant or recessive. If it is dominant, given the definitions below, identify whether it is caused by haploinsufficiency, gain of function, or a dominant negative phenotype.

   Definitions of types of dominance:
   Haploinsufficiency – One dose of the gene product of the wildtype allele is too little to produce a wildtype phenotype.
   Gain of function – A mutant phenotype is produced because the disease gene has a new function. It is very unusual for the function to be completely novel. Most often, the gene is expressed at a time, in a tissue, or in at an elevated level that is unusual or the regulation of the protein is abnormal.
   Dominant negative – The mutant protein interferes with the function of proteins from the wildtype allele. This is most common for proteins that multimerize to form dimers or more complex structures.

   a. In Huntington's disease, there is an expansion of a repeated CAGCAGCAGCAG region within the gene so that there are 36 to over 100 repeats of CAG, resulting in the inclusion of 36 to over 100 glutamine amino acids in the huntingtin protein. (The wildtype protein contains 10-28 repeats of glutamine). The mutant proteins do not fold into functional proteins, and instead aggregate with other mutant proteins, which has a toxic effect in some cells. This causes loss of brain tissue, behavioral changes, dementia, and uncontrolled muscle movements. Other mutations that interfere with the function of the huntingtin protein do not cause Huntington's disease or any known phenotype if the mutations are present in only one allele.

Class Activity 2A: The Role of Proteins in Genetic Disorders

b. Achondroplasia is caused by a specific mutation in the *FGFR3* gene (Fibroblast Growth Factor Receptor 3) that encodes a cell-surface receptor protein. The function of the wildtype protein is to interact with growth factors outside of the cells that might form bone and to send signals inside the cell that inhibit the conversion of cartilage tissue into bone tissue. In the mutant FGFR3 protein, glycine at amino acid position #380 is replaced by an arginine. This mutation causes the receptor to be overactive, interfering with skeletal development. This results in short-limbed dwarfism.

c. Osteogenesis imperfecta type I is caused by mutations in a gene for type I collagen, a protein that is abundant in bone and that normally gives bone some flexibility. Collagen forms rigid fibers that play a structural role in bones, tendons, skin, and teeth. Collagen is the most abundant protein in the human body. The mutations in the collagen gene in osteogenesis imperfecta type I cause less collagen than normal to be produced, resulting in brittle bones that are prone to fracture.

d. Osteogenesis imperfecta types II, III, and IV are cause by mutations in either the gene for type I collagen or for type II collagen. Collagen proteins are abundant in bone and that normally gives bone some flexibility. Three collagen polypeptides interact by packing together to form a triple helix that is very rigid in the mature, quaternary structure of collagen. Mutations in type I or type II collagen in osteogenesis imperfecta types II, III, and IV are mutations that interfere with the ability of the collagen polypeptide to pack properly into a rigid triple helix. This means that any collagen triple helix containing any mutant collagen will not be functional. This results in brittle bones that are prone to fracture.

collagen triple-helix (1CAG.pdb)
(30 of about 1400 amino acids are shown here in each of the 3 collagen polypeptides)

e. Osteogenesis imperfecta types VII and VIII are caused by mutations in the genes *CRTAP* and *LEPRE1*, respectively. The proteins produced by these genes (cartilage–associated protein and leprecan) work together in an enzyme complex that processes immature collagen polypeptides into a mature form. When translated, a collagen polypeptide has a repeating pattern of the amino acids glycine-proline-proline. The Crtap/leprecan enzyme complex makes collagen mature by adding a hydroxide group to half of the proline amino acids in collagen (a cofactor in this reaction is vitamin C; deficiency of vitamin C causes scurvy due to too little of this reaction). This formation of hydroxyproline is required for proper collagen function. The absence of either Crtap or leprecan prevents the production of normal collagen, resulting in a very severe form of osteogenesis imperfecta in which bones are very brittle and very prone to fracture.

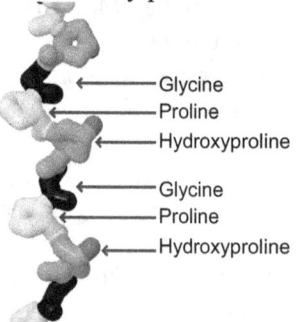

Presence of hydroxyproline in collagen (1CAG.pdb)

It's necessary to study each mutation individually to determine its effect with certainty.

8. Why does the mechanism of the mutation in causing disease matter? Consider how it might impact strategies for treatment of the disease. In which cases, should you try to introduce more of the gene or its protein product? In which cases should you try to turn off the mutant gene?

9. If two people have genetic diseases with the same phenotype, can you conclude that they have mutations in the same gene? Explain.

10. Based on the examples above (and your reasoning skills), do you think that recessive loss of function mutations are more likely to occur in enzymes or in proteins that serve a structural function? Why?

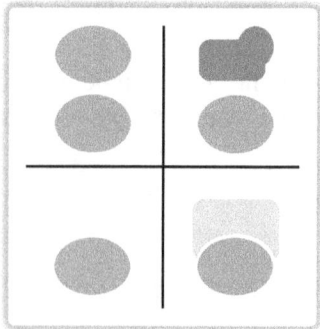

Chapter 2 Reading #2 and Discussion Questions

Reading: *Molecular Biology: Concepts for Inquiry*
Chapter 2: Section 2.1B

Online Quiz: 2.1B

Discussion Questions:

1. Do "dominant" and "recessive" imply how frequently a mutation is observed in a population? Explain.

2. Two parents are both homozygous for the wildtype allele of a particular gene. They have a child who has a genetic disorder due to a new mutation that eliminates the function of one allele of the gene. Does the child have a dominant or a recessive disorder? Explain how you know.

3. Compare and contrast a dominant negative mutation with a loss of function mutation and a gain of function mutation.

4. If a doctor sees a new patient in the clinic who has a phenotype consistent with having a genetic disorder, what information would the doctor need to gather to determine the mode of inheritance and type of mutation for this patient's disease?

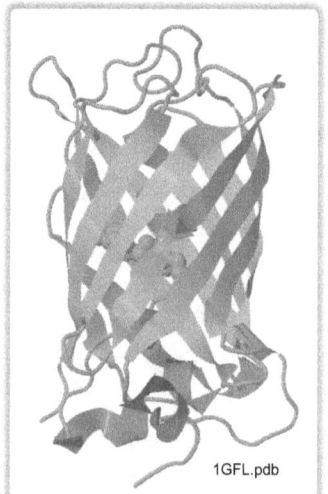

1GFL.pdb

Chapter 2 Reading #3 and Discussion Questions

Reading: *Molecular Biology: Concepts for Inquiry*
Chapter 2: Sections 2.1C-F

Online Quiz: 2.1C-F

Discussion Questions:

1. What are protein secondary structures? How do they form?

2. How do protein tertiary and quaternary structures form? What roles do amino acid sidechains play in protein folding? In what locations within a protein are different classes of sidechains usually found?

3. What role can disulfide bonds play in the formation of stable quaternary structures? Provide an example.

4. What is the role of the formation of intramolecular electrostatic attractions in the spontaneity of protein folding? What is the role of the hydrophobic effect in the spontaneity of protein folding?

5. Explain why it is not correct to say that "hydrophobic molecules stick to hydrophobic molecules?" Why do you think that from a certain perspective this statement is not completely incorrect? Do you think taking this particular shortcut in reasoning is unwise? Can you think of any situation in which it would be problematic?

6. How does Figure 2.15 illustrate the thermodynamics of protein folding?

7. What are the benefits of chaperone/chaperonin proteins?

8. What factors might promote protein misfolding? Provide an example of how protein misfolding can cause disease (other than prion proteins).

9. Explain how prion proteins cause disease. Why are prion proteins considered infectious?

10. When prion proteins were first studied, the scientific community was largely skeptical that proteins could be the infectious particles in prion diseases. Why was this skepticism understandable? What concepts do you think might have shaped the skeptics' thinking?

11. Why do cells need proteasomes?

12. How are Alzheimer's disease and prion diseases similar? How are they different?

13. How does understanding concepts related to protein folding, protein misfolding, and the processing of misfolded proteins inform your thinking about biological processes underlying Alzheimer's disease?

Class Activity 2B: Fluorescent Protein Analysis

Work with others in your small class group to complete these questions. Thoroughly discuss one question at a time without skipping ahead. Use your logic skills to answer the questions - do not look up answers in your textbook or in another reference.

Color photos showing actual data that was the basis for this activity are provided on https://hackettmolecularbiology.blogspot.com/ and it is recommended that you view those data along with the drawings provided here.

Fluorescent proteins are very frequently used in the lab as a visual output in experiments. Green fluorescent protein (GFP) was discovered in the jellyfish *Aequorea victoria*. It has been cloned into bacteria and has been mutated to make multiple other colors, including blue fluorescent protein (BFP) and yellow fluorescent protein (YFP). Separately, a red fluorescent protein (DsRed) was discovered in the coral *Discosoma*. DsRed naturally exists as a tetramer, so it was mutated at more than 30 different amino acids to allow it to act as a monomer when expressed in mammalian cells, creating monomeric red fluorescent protein (RFP). RFP has also been mutated to produce a variety of different colors. Roger Tsien's lab produced many of these and named them after fruits, including mTangerine (red-orange), mCherry (red), and mGrape (purple). You can view photos of bacteria expressing BFP, GFP, YFP, mTangerine, mCherry, and mGrape at https://hackettmolecularbiology.blogspot.com/. Notice that with the exception of BFP, the color of these proteins is visible under normal room light. However, these proteins fluoresce when exposed to UV or blue light.

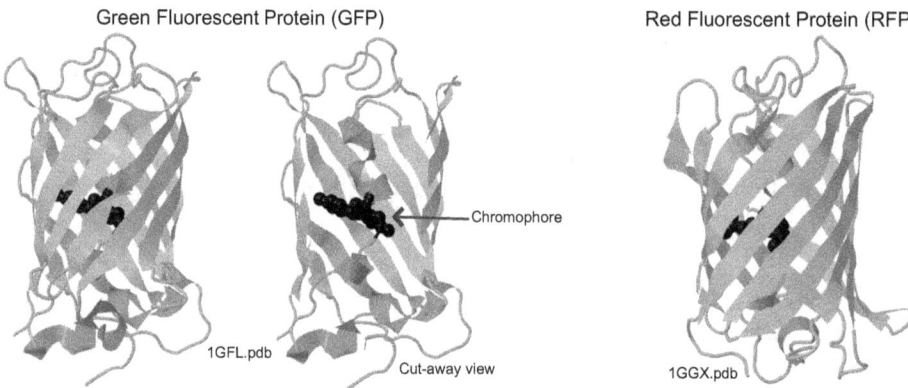

GFP and RFP both have similar beta-barrel shapes and are about the same mass. In the center of the barrels of these proteins is a chromophore, a molecular group that formed from three adjacent amino acids. The chromophore absorbs higher energy light (shorter wavelength) to take on an excited state. Then when the chromophore returns to the ground state, it emits light of a lower energy (longer wavelength). Some of the energy is used up in that process.

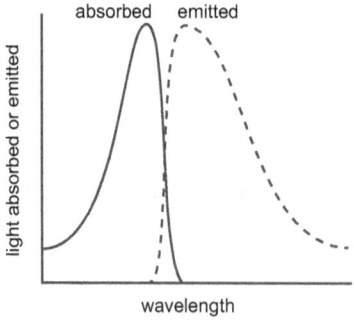

Fluorescence occurs when a molecule absorbs one wavelength of light and then emits a different wavelength of light. The following table summarizes the wavelengths of light that fluorescent proteins absorb and emit. Notice that since energy can't be created from nothing, the absorbed light is always of higher energy (shorter wavelength) than the emitted light. Also note that the given wavelengths are for the peak absorption and emission, but a range of wavelengths of light can be absorbed and a range of wavelengths is emitted as shown in the graph above.

Source	Fluorescent Protein	Peak Wavelength Absorbed	Peak Wavelength Emitted
GFP-derived	BFP	380 nm	440 nm
	GFP	488 nm	509 nm
	YFP	516 nm	529 nm
RFP-derived	mTangerine	568 nm	585 nm
	mCherry	587 nm	596 nm
	mGrape1	595 nm	620 nm

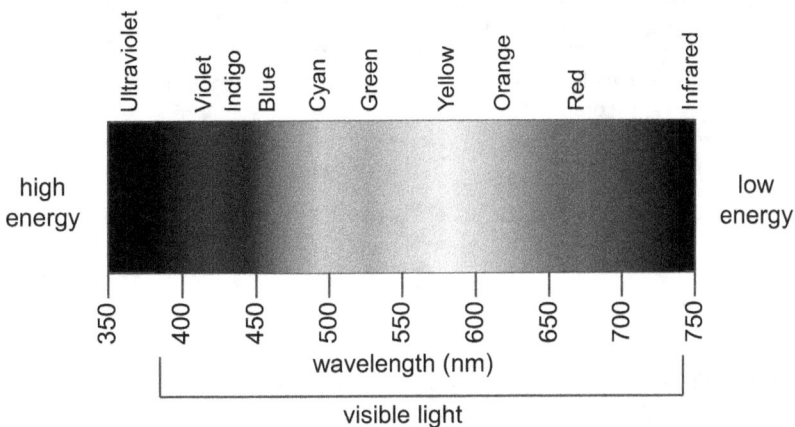

1. The photos below show patches of bacteria expressing fluorescent proteins. Explain why BFP is brighter than YFP in the left-hand photo taken in UV light, but YFP is brighter than BFP in the right-hand photo taken in blue light.

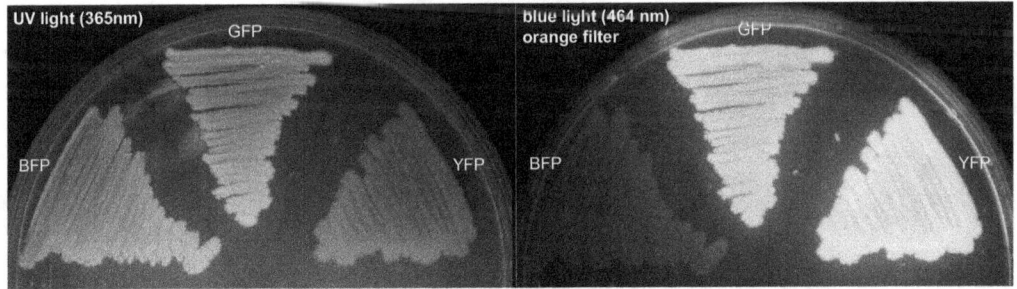

Students conducting the BioBridge experiments developed for fluorescent proteins (Princeton Molecular Biology Outreach) purified fluorescent proteins from bacteria. The fluorescent protein solutions were mixed with a gel loading buffer and then were immediately loaded into wells on an agarose gel. A prestained protein ladder where dyes were attached to the proteins was also run on the gel. Cartoons summarizing the data and a panel showing one of the original gels are provided below. Color photos showing multiple examples of student data are provided on https://hackettmolecularbiology.blogspot.com/.

Class Activity 2B: Fluorescent Protein Analysis

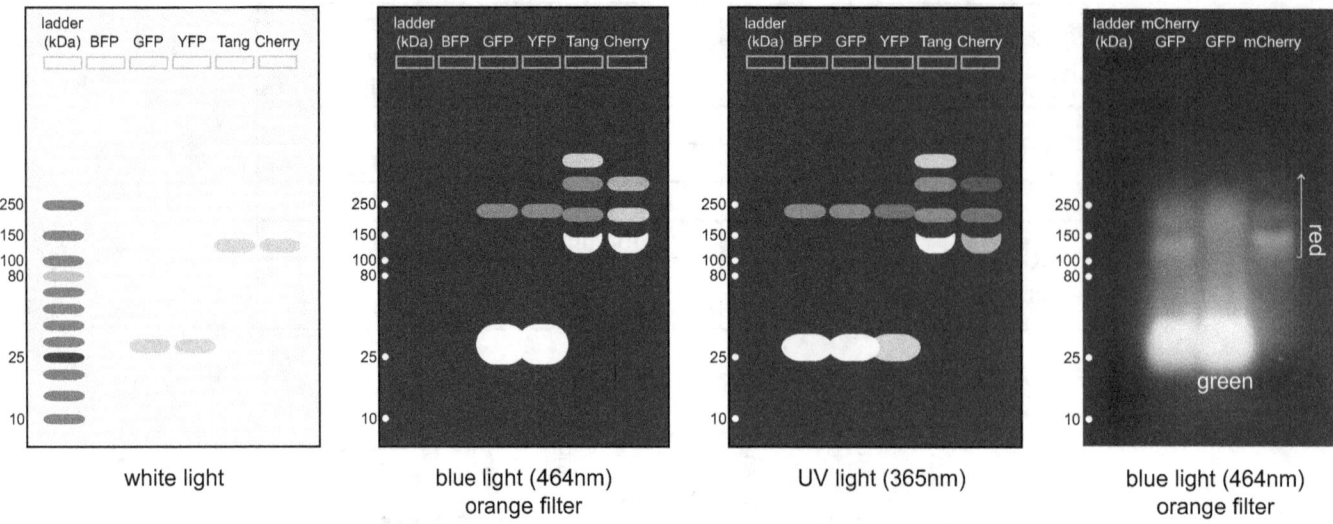

2. How do proteins closer to the wells differ from proteins further from the wells?

3. Why is the ladder only visible under white light? Why are the fluorescent proteins most visible under blue light or UV light? Why is BFP only visible under UV light?

4. GFP and RFP have similar masses, 26.2 kDa for GFP and 27 kDa for RFP. The other fluorescent proteins are missense point mutants of GFP and RFP. With this in mind, explain in what respects the data above are unexpected.

5. Hypothesize about what might explain these unexpected data.

6. What kind of experiment might be conducted to test your hypothesis?

STOP-check with your instructor before moving on.

38 | Unit 2   *Class Activity 2B: Fluorescent Protein Analysis*

7. Were the fluorescent proteins denatured on the above agarose gels? How do you know?

8. Proteins are not usually run on agarose gels (you can see in the agarose gel image that the bands were very fuzzy and the proteins seemed to be diffusing a significant amount). Polyacrylamide gels provide much better resolution. Additionally, protein samples are typically boiled after being combined with SDS loading buffer, but before being loaded on the gel. Samples were not boiled before being loaded on the above agarose gel. Why do you think it's generally standard practice to boil samples before loading?

9. When fluorescent proteins were run on a polyacrylamide gel, boiling samples substantially reduced the level of large bands compared to when samples were run on a polyacrylamide gel without being boiled. Why does this make sense? Do these data support your earlier hypothesis to explain the agarose gel results?

10. The agarose gel data above is evidence that under some circumstances proteins can aggregate. Describe a disease that results from protein aggregation. What are ways that cells attempt to prevent protein aggregation?

11. What kind of protein secondary structure is most associated with aggregation? Is this also a feature of fluorescent proteins?

12. Why do each of these cause protein denaturation: heat? acid? base? detergent?

13. It's easy to observe when fluorescent proteins denature – how? A student tries to denature BFP, GFP, and YFP using different combinations of heat, acid, base, and detergent. The conditions under which these proteins denature varies somewhat. What does this suggest about the effect of point mutations on protein structure? Do the point mutations that created BFP (Y66H) and YFP (T203Y) also affect protein function? How do you know?

Reference: http://hhmi.princeton.edu/images/documents/lab_protocols/HHMI_2013_Lab_5_Fluor_Proteins-2.pdf

Chapter 2 Reading #4 and Discussion Questions

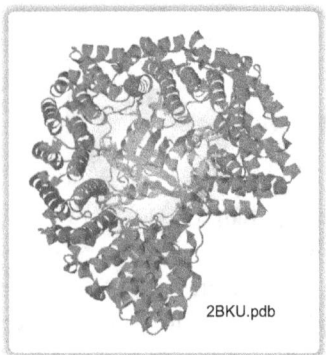

Reading: *Molecular Biology: Concepts for Inquiry*
Chapter 2: Sections 2.1G-I

Online Quiz: 2.1G-I

Discussion Questions:

1. What are the advantages and disadvantages of different methods for determining/predicting protein structure?

2. How does the synthesis of transmembrane proteins differ from the synthesis of cytoplasmic proteins?

3. Which parts of the cell are considered to be spatially contiguous with the extracellular space? Why? Explain how this is relevant to protein synthesis and transport. For example, consider how different portions of a transmembrane protein will be situated at different times in its synthesis and transport.

4. Explain the thermodynamic forces that drive the spontaneous interaction of a protein with another molecule.

5. How does the lock and key model differ from the induced fit model? Which is more correct?

6. If a protein binds another molecule, do they stay bound forever? Explain. (What do we mean by "bind" here?) What role do you think understanding the kinetics of these interactions might play in drug design?

Chapter 2 Reading #5 and Discussion Questions

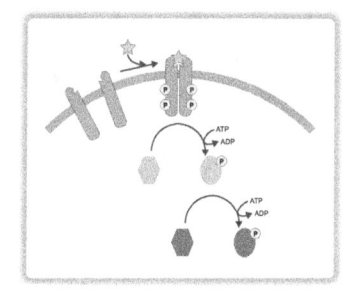

Reading: *Molecular Biology: Concepts for Inquiry*
           Chapter 2: Sections 2.1J-K, Sections 2.2A-D

Online Quiz: 2.1J-K, 2.2A-D

Discussion Questions:

1. Explain the role of phosphorylation in determining protein shape. How is this related to protein function?

2. Why does it make sense that kinases are frequently found in signaling cascades?

3. Why do cells need phosphatases?

4. How does protein ubiquitination help to prevent disease?

5. How does a change in temperature change enzyme activity? How does a change in pH change enzyme activity?

6. How do enzymes speed up chemical reactions?

7. What role do enzymes play in increasing the spontaneity of processes in the cell? What factors influencing spontaneity are unaffected by enzymes?

8. Why do people need to eat vitamins? Minerals?

9. How might EDTA inhibit the function of an enzyme?

10. What are proteases? How are they regulated? Why is this necessary?

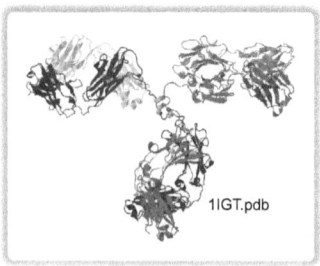

Chapter 2 Reading #6 and Discussion Questions

Reading: *Molecular Biology: Concepts for Inquiry*
Chapter 2: Section 2.3A-C

Online Quiz: 2.3A-C

Discussion Questions:

1. What general principles should be kept in mind when designing a drug?

2. Why do you think that proteins are much more frequently drug targets than DNA or RNA?

3. What role do the chemical properties of proteins play in their purification?

4. Provide an example of how protein purification is necessary for treatment of a genetic disease. Is it a loss-of-function or gain-of-function disease? Why does this make sense?

5. What are "immobilized" enzymes? What benefit do they have?

6. How can antibodies be used to detect proteins? Will an antibody that recognizes hemoglobin also recognize collagen? Explain.

7. In what situations are fluorescent proteins useful for detecting other proteins? What must be done in order to make this possible?

Unit 2 Self-Assessment

1. Hereditary nephrogenic diabetes insidious is a disorder in which the kidney tubules remove too much water from the blood. The patient produces too much urine and is excessively thirsty. Antidiuretic hormone (ADH) normally limits urine production by signaling to the kidney tubule cells to insert more water channels (Aqp2 proteins) in their membranes, resulting in more water being returned from the tubules to the blood. However, in one form of this disorder, mutant Aqp2 protein can't be inserted into the membrane properly and mutant Aqp2 protein binds to wildtype Aqp2 protein, limiting the insertion of these water channels into the tubule cell membranes.
    a. Would you predict that the disorder is dominant or recessive?
       Explain why you made this choice.

    b. Which of the following best describes the type of disorder: recessive loss of function, haploinsufficiency, gain of function, or dominant negative? Explain.

2. If a portion of a transmembrane protein is located in the lumen of the rough ER following protein synthesis, will that portion of the protein end up inside or outside of the cell membrane?

3. For the following DNA sequence, assume the top strand is the sense strand. State the sequence of (a) the mRNA and (b) the polypeptide that would be produced from this DNA. (Assume transcription and translation both begin at the first nucleotide).
       5'ATGCCCAGCCTGGGATGACGA 3'
       3'TACGGGTCGGACCCTACTGCT 5'

4. State two ways that a genetic engineer would need to alter a eukaryotic gene before being able to express the gene in bacteria.

5. You are a researcher who is trying to learn more about the mechanism of a dominant disease. You aren't sure which gene causes the disease, but have mapped the location of the gene to a small chromosomal region that contains two genes, A and B. You sequence both genes in several affected and unaffected members of one family and discover that the sequences of each of these genes is identical in all of the family members except for the following differences (sense strand listed):

	Unaffected family members	Affected family members
gene A codon #65	allele 1: CAC allele 2: CAC	allele 1: CAT allele 2: CAC
gene B codon #5	allele 1: AGA allele 2: AGA	allele 1: TGA allele 2: AGA

    a. Mutations in which gene most likely cause the disease? Explain.

    b. To treat the disease, should doctors try to turn off the gene, add more wildtype protein, or both? Explain.

    c. Does the disease gene more likely code for an enzyme or a protein that's not an enzyme?

6. Explain the role of the coupling of spontaneous and nonspontaneous processes in either transcription or translation.

7. State one strategy cells use to prevent the aggregation of proteins.

8. Thermodynamically, why is it difficult or impossible for cells to make amyloid-like proteins that have misfolded and form aggregated fibrils refold into their normal, native structures?

9. Place a check mark in each box that describes the chemistry of the sidechain of each amino acid.

amino acid	hydrophilic	hydrophobic	neutral	acidic	basic
phenylalanine					
serine					
leucine					
arginine					

10. a. Name the two main kinds of secondary structures that form due to hydrogen bonding between atoms in the backbone of proteins.

    b. Briefly state the 4 "rules" that govern the formation of protein tertiary structure.

    c. Choose one of the "rules" you listed above and describe how it contributes to the spontaneity of protein folding.

11. At the molecular level, how do enzymes lower the activation energy in a reaction?

12. A researcher purifies the zymogen form of the stomach protease pepsin. The pH of the stomach is very acidic. State two steps the researcher will need to take before the pepsin is active.

13. a. Explain why it makes sense that abnormally folded prion proteins often have structures where beta sheets are more prominent than alpha helices.

    b. Excluding an increased likelihood of exposure to infectious prions, state one other factor that might make a person's chance of developing a prion disease be higher than average.

14. The hydrophobic effect contributes to the spontaneity of protein folding primarily because:
    a. Fewer H-bonds are formed, increasing ΔH.
    b. Fewer H-bonds are formed, decreasing ΔH.
    c. More H-bonds are formed, increasing ΔH.
    d. More H-bonds are formed, decreasing ΔH.

15. Circle ALL of the following that might INCREASE the activity of an enzyme.
    a. Decreasing the temperature by 5°C
    b. Increasing the temperature by 5°C
    c. Decreasing the pH
    d. Increasing the pH
    e. Adding a drug that competes with substrate for binding in the active site.
    f. Adding magnesium ions
    g. Adding EDTA
    h. Mutating one amino acid in the enzyme
    i. Boiling the enzyme
    j. Phosphorylating the enzyme

Unit 3

DNA Replication, Repair and Genetic Engineering

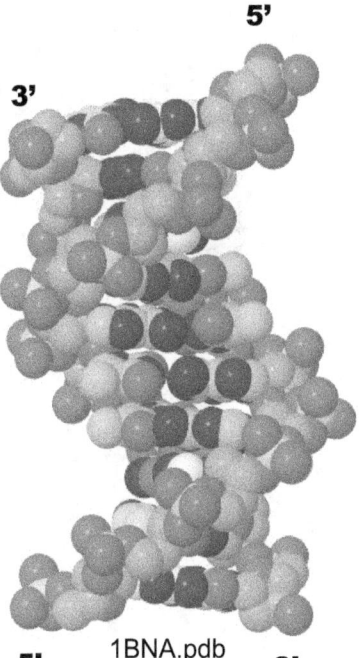
1BNA.pdb

From an evolutionary perspective, the purpose of an organism and all of the cells and proteins it contains is to pass on its DNA to the next generation. Of course, most of us would argue that there's much more to being human than just passing on our DNA, but in this unit we'll focus on the strategies organisms use to faithfully pass on their hereditary material, their DNA. In the previous unit, you learned how DNA codes for proteins. In this unit, you will learn how the functions of some of those proteins are to copy, modify, and repair DNA. We will then apply this understanding to an exploration of how DNA modifying enzymes are exploited in vitro and in vivo for the purpose of genetic engineering.

To achieve the deepest level of understanding, you are encouraged to complete the class activities, textbook readings and question sets in the order listed below.

Suggested order:
Chapter 3 Reading #1 and Discussion Questions
Chapter 3 Reading #2 and Discussion Questions
Class Activity 3A: Inquiry into PCR
Chapter 3 Reading #3 and Discussion Questions
[Experiment: Detection of Genetically Modified Food Through PCR]
Chapter 3 Reading #4 and Discussion Questions
Chapter 3 Reading #5 and Discussion Questions
Class Activity 3B: Restriction Enzymes and Cloning
Class Activity 3C: Gibson Assembly Mutagenesis of Fluorescent Protein Genes
[Experiment: Gibson Assembly Mutagenesis of Fluorescent Protein Genes]
Unit 3 Self-Assessment Questions

Prior knowledge (review if necessary):
- Structure of DNA, RNA, and proteins (in Chapters 1, 2)
- Complementary DNA and RNA nucleotides (in Chapter 1)
- Coupling of spontaneous processes to nonspontaneous processes (in Chapter 1)
- Thermodynamic principles: free energy, enthalpy, entropy (in Chapter 1)
- How DNA codes for RNA which codes for protein (in Chapter 2)
- How mutations contribute to dominant and recessive phenotypes (in Chapter 2)
- Oligomeric proteins (in Chapter 2)
- Factors influencing the binding of proteins to other molecules (in Chapter 2)
- Enzyme structure and function (in Chapter 2)

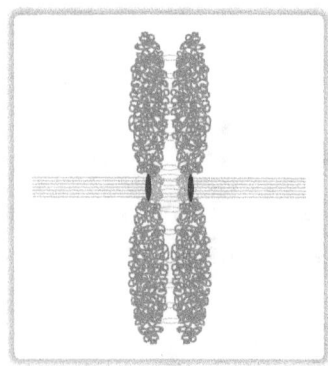

Chapter 3 Reading #1 and Discussion Questions

Reading: *Molecular Biology: Concepts for Inquiry*
Chapter 3: Introduction and Section 3.1

Online Quiz: 3.1

Discussion Questions:

1. Explain the relationship between DNA, genes, and chromosomes.

2. The drug Taxol stabilizes tubulin oligomers and protects them from disassembly. What effect do you think this drug would have on mitosis? Given that cancer cells divide more frequently than most normal cells, why do you think Taxol is effective as a chemotherapy treatment for cancer?

3. Explain all of the ways in which mitosis would not happen properly if cohesin proteins were not present in the cell.

4. Explain the function of centromeres. Why is it not accurate to say that the centromere is the middle of a chromosome?

5. Review the methods for gene transfer in prokaryotes (Fig. 3.7). Why do you think transformation is used more commonly than transduction and transduction is used more commonly than conjugation by scientists putting DNA into bacteria for genetic engineering purposes?

6. What two sequences must always be part of a bacterial plasmid used for genetic engineering? For each of these, explain what you think the consequences would be if the plasmid did not contain that component.

Chapter 3 Reading #2 and Discussion Questions

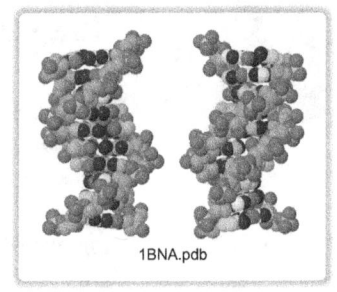

Reading: *Molecular Biology: Concepts for Inquiry*
           Chapter 3: Section 3.2

Online Quiz: 3.2

Discussion Questions:

1. Some proteins that scan the DNA for new mutations recognize mismatches between DNA nucleotides because they create either a bulge or a narrowing of the DNA double helix. Which nucleotide pairs would create a bulge? A narrowing?

2. After observing the bottom panel of Figure 3.10, why does it make sense that most proteins that recognize a specific DNA sequence bind in the major groove?

3. How many bp are in 1.3 kb? How many nt?

4. How does the structure of DNA allow it to migrate on an agarose gel? What is the basis of the separation of a DNA solution into multiple bands on a gel?

5. Based on the structure of DNA, what type of sidechains do you think should be used by a protein to bind DNA?

6. If a protein uses zinc fingers to specifically recognize 12bp of DNA, how many zinc fingers does it contain?

Class Activity 3A: Inquiry into PCR

Work with others in your small class group to complete these questions. Thoroughly discuss one question at a time without skipping ahead. Use your logic skills to answer the questions - do not look up answers in your textbook or in another reference.

1. Study the mechanism of DNA replication:

   a) Double-stranded DNA

   b) Proteins bind the origin of replication and separate the DNA strands (not shown). Helicase separates the strands on either side of the origin. Primase synthesizes a short RNA primer in the 5' to 3' direction that is complementary to the DNA.

   c) DNA polymerase can only link the 5' phosphate group of a new nucleotide to the 3' OH of an existing nucleotide. Thus, DNA polymerase can't add nucleotides anywhere indiscriminately, but requires a primer to start at a particular position. Starting at the 3' end of an RNA primer, DNA polymerase adds complementary DNA nucleotides in a 5' to 3' direction. When DNA polymerase encounters the 5' end of the next RNA primer, it removes the RNA nucleotides and adds DNA nucleotides.

   d) DNA polymerase can only add nucleotides in a 5' to 3' direction. The leading strand is synthesized continuously as DNA polymerase moves forward. Since DNA polymerase can't add nucleotides backwards the lagging strand is synthesized one short segment at a time.

   e) The short DNA segments on the lagging strand are called Okazaki fragments. Gaps remain between the fragments because DNA polymerase could link the last nucleotide it inserted to the 3' end of the previous nucleotide, but it couldn't link it to the 5' end of the next nucleotide. DNA ligase seals the gaps by forming the final covalent phosphodiester bonds between nucleotides at these locations. DNA ligase actually follows DNA polymerase and does this immediately, but this is depicted as a separate step here for clarity.

   f) Two copies of the same double-stranded DNA molecule. Replication is semi-conservative because the new double helix consists of one old strand (blue) and one new strand (magenta).

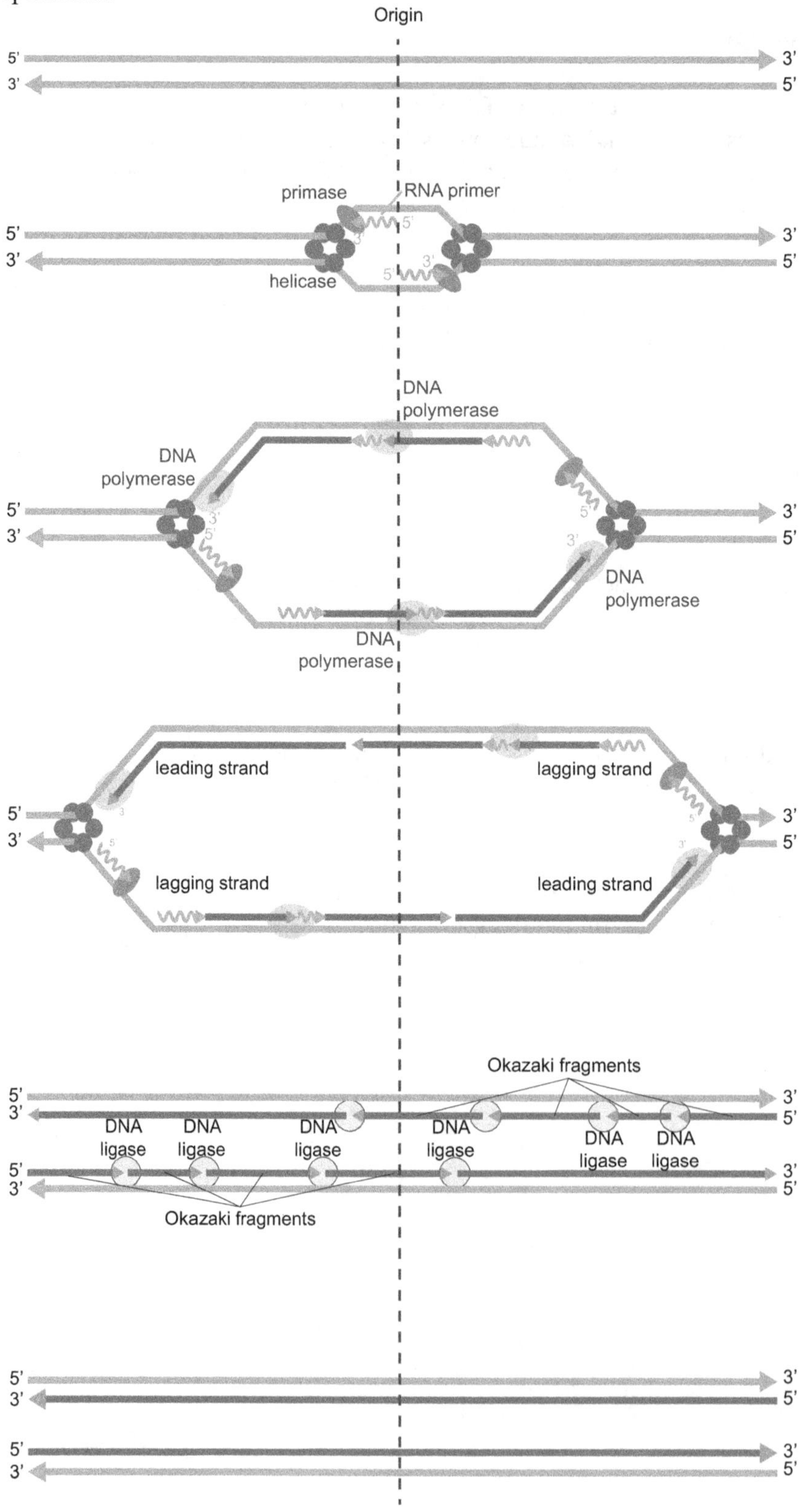

2. Identify the two processes that must take place before DNA polymerase can copy DNA.

Process	Enzyme used
1.	
2.	

3. Can DNA polymerase start copying the DNA anywhere in the genome? Explain.

4. The annealing of two complementary DNA strands to form double-stranded DNA is sometimes a spontaneous process in cells. Explain in terms of ΔH, ΔS and ΔG. For each, explain why the value changes in this direction. Under what conditions is it spontaneous?

5. The polymerase chain reaction (PCR) is a method for carrying out DNA replication of a particular segment of DNA *in vitro* in a plastic tube. Double-stranded DNA that contains the sequence to be copied is added to the tube. **The only enzyme that is added is DNA polymerase.**
    a. Since helicase is NOT added, what do you think could be done instead to create the same effect?

    b. You only want to copy a particular segment of the DNA. In addition, DNA polymerase can't start copying just anywhere. What kind of molecule do you think you would need to add to specify where DNA polymerase starts copying?

6. As discussed in question 4, two single strands of DNA can anneal to form double-stranded DNA if the energy released as the new hydrogen bonds form is more significant than the unfavorable TΔS term.

   a. a. Are longer or shorter DNA segments more likely to anneal at a given temperature? Why?

   b. Are G-C or A-T base pairs more favorable to annealing? Why?

7. The molecule you described in 5b is called a primer. It's a short single-stranded DNA molecule with a custom sequence that can be ordered from a company. It's added to the PCR tube at a very high molarity relative to the template DNA. (You might have written "RNA primer" above, but if you just make the primer from DNA, the RNA nucleotides won't need to be replaced later).

   a. If the first step in PCR is to raise the temperature to 95°C to denature the DNA (separating the strands), what would you need to do to allow a primer to anneal?

   b. How could you control at what temperature the primer anneals?

8. What's the problem with having DNA polymerase in the same tube that gets heated to 95°C to denature the DNA?

9. To solve the problem you identified in #8, researchers decided to use a DNA polymerase isolated from bacteria that live in hot springs, *Thermus aquaticus*. This enzyme, called *Taq* polymerase doesn't denature at 95°C and has optimal activity at 72°C. At what temperature would you put a PCR tube to allow DNA polymerase to copy the DNA?

10. Complete the following chart to identify the steps in PCR:

Step	What happens	Temperature
Denaturation		
Annealing		A temperature below 72°C
Elongation		

11. To copy a specific segment of DNA, **two** primers are used in a PCR reaction. Why do you think this is necessary?

12. List everything that you think would need to be in the tube for a PCR reaction.

13. If the PCR reaction mix is put into a thermal cycler (a machine that controls the temperature) and if the temperature changes listed in #10 are repeated 30 times, how many copies of the specific DNA segment will be made?

Class Activity 3A: Inquiry into PCR

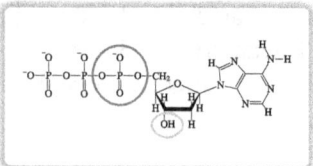

Chapter 3 Reading #3 and Discussion Questions

Reading: *Molecular Biology: Concepts for Inquiry*
Chapter 3: Section 3.3

Online Quiz: 3.3

Discussion Questions:

1. Would you predict that a bacterial chromosome or a human chromosome would have more origins of replication? Explain why.

2. Why are Okazaki fragments a necessary part of DNA replication?

3. Why do you think the enzyme primase evolved and that its function is not a function of DNA polymerase?

4. What is nonspontaneous about DNA replication? What makes the process spontaneous overall?

5. Compare and contrast PCR with *in vivo* DNA replication. (As part of this, for each part of *in vivo* replication, specify the analogous process in PCR).

6. How did each of the following contribute to the development of PCR? A. Knowledge of DNA structure. B. Knowledge of the effect of environmental conditions on the evolution of protein structure. C. Understanding of the principles of thermodynamics.

7. What is the goal of any PCR reaction? Describe some specific applications of PCR.

8. Compare and contrast PCR and DNA sequencing.

9. Explain the function of dideoxynucleotides in DNA sequencing.

10. Explain how the design of the Human Genome Project contributed to the accuracy of the assembled genome sequence.

Chapter 3 Reading #4 and Discussion Questions

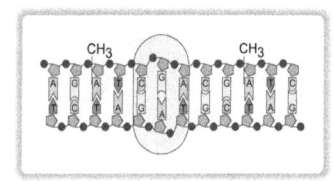

Reading: *Molecular Biology: Concepts for Inquiry*
Chapter 3: Section 3.4

Online Quiz: 3.4

Discussion Questions:

1. How are the rate of DNA damage, the rate of DNA repair, the rate of apoptosis, and mutation rate related to each other?

2. How does a cell recognize that its DNA has been damaged?

3. State the roles of the following in the repair of point mutations: endonucleases, exonucleases, DNA polymerase, DNA ligase.

4. During mismatch repair, explain how the repair proteins know which nucleotide within a mismatched base pair is the mistake. At what times during the cell cycle is mismatch repair most likely to correctly repair the DNA?

5. What kinds of factors tend to cause point mutations? What kinds of factors tend to cause DNA breaks?

6. Compare the circumstances under which cells might preferentially use nonhomologous end-joining or homologous recombination for the repair of double-strand breaks.

7. Explain the role of some of the DNA replication machinery in homologous recombination.

8. Why might DNA mismatch repair proteins be needed to complete homologous recombination?

9. What do genetic disorders in which DNA repair proteins are mutated have in common?

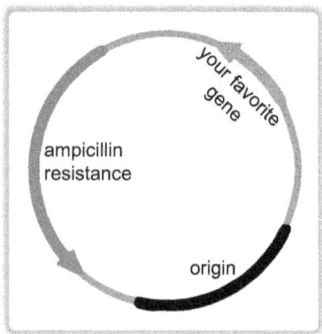

Chapter 3 Reading #5 and Discussion Questions

Reading: *Molecular Biology: Concepts for Inquiry*
Chapter 3: Section 3.5

Online Quiz: 3.5

Discussion Questions:

1. What do restriction enzymes do? How do they do it? What prevents them from doing this to the DNA in the cells where they are made?

2. To what does the "restriction" part of the name "restriction enzyme" refer? Explain how experiments led to the discovery of restriction enzymes.

3. Compare and contrast how the restriction enzymes *Eco*RI and *Eco*RV bind to and cut DNA.

4. Given the way that restriction enzymes recognize their cut site and make their cuts, why does it make sense that many are dimers?

5. What are the "scissors" and "glue" used in genetic engineering?

6. You cut two different plasmids with restriction enzymes that cut each plasmid into two pieces. You want to combine one fragment from each digest to create a new plasmid. Why can't you simply mix the two restriction digests together, add ligase, and transform the DNA? What would you need to do instead and why?

7. Which would you prefer to use in a genetic engineering project and why: a restriction enzyme that leaves sticky ends or a restriction enzyme that produces blunt ends?

8. Explain the factors that contribute to the efficiency of ligation reactions.

9. How can you ensure that your ligation reaction products contain more plasmids that receive an insert than plasmids that don't receive an insert?

10. What should be considered when cloning a PCR product into a plasmid?

11. Explain how the enzyme reverse transcriptase is a useful tool in molecular biology research.

12. After performing a ligation reaction, what must be done in order to isolate DNA containing the correctly ligated plasmid?

13. Compare/contrast transient transfection, stable transfection, transduction, and homologous recombination as methods for introducing DNA into eukaryotic cells: Consider the ease of use, whether or not the DNA integrates into the genome, if it integrates does it do so at a specific site?, and do you think it would be a good technique for introducing DNA into human patients – why?

14. Explain how homologous recombination could be used to produce either a transgenic mouse or a knockout mouse.

15. Compare the advantages of using PCR or Southern blotting to detect an insertion into genomic DNA.

16. Cystic fibrosis is a recessive genetic disease where airways become clogged with mucus due to the loss of function of the CFTR protein that normally transports chlorine ions across the cell membrane. You want to insert a wildtype copy of the *CFTR* gene into the genome of a patient with cystic fibrosis to cure the disease.
    a. Explain everything that is wrong with this strategy: To do so, you decide to PCR a wildtype copy of *CFTR* using primers that have tails containing *Eco*RI restriction enzyme sites. You cut a sample of the patient's genomic DNA with *Eco*RI and cut the *CFTR* PCR product with *Eco*RI. You mix these digests together and add DNA ligase.
    b. What would be a better strategy?

Class Activity 3B: Restriction Enzymes and Cloning

Work with others in your small class group to complete these questions. Thoroughly discuss one question at a time without skipping ahead. Use your logic skills to answer the questions - do not look up answers in your textbook or in another reference.

1. Which of the following is most likely to be a restriction enzyme recognition site? Why?

   a. TCCAGG   b. GGATCC   c. GATCGA   d. GGGGGG

2. Look up the restriction enzymes *Sma*I and *Xma*I in Appendix 4 of this workbook.
   a. Compare and contrast how they recognize and cut DNA.

   b. Speculate why a researcher might choose to use one of these instead of the other in their genetic engineering.

3. Consider the restriction enzymes described in Appendix 4.
   a. Most restriction enzymes recognize a palindromic sequence and cut within that sequence. Name and give the recognition/cut site for a restriction enzyme (not listed above) that does this and that creates:
      i. Sticky ends

      ii. Blunt ends

   b. Name and give the recognition/cut site for a restriction enzyme that recognizes an asymmetric, nonpalindromic sequence and that cuts outside of that sequence.

   c. How do you think the structure of the enzymes you listed in (a) might differ from the structure of the enzyme listed in (b)?

4. Which of the following are good cloning strategies to put gene A into plasmid B? What's wrong with the others? Assume that gene A itself does not contain either *Eco*RI or *Bam*HI cut sites. Assume that plasmid B has only one *Eco*RI site and only one *Bam*HI site. These sites are near each other and not near *Amp*^R or the origin. Assume sites are present where the questions imply. Assume all desired fragments have been gel-purified.

   a. Cut gene A out of plasmid A using *Eco*RI. Digest plasmid B with *Eco*RI. Ligate the gene A fragment into plasmid B. Transform.

   b. PCR gene A using primers with *Eco*RI sites in their tails. Cut the PCR product with *Eco*RI. Digest plasmid B with *Eco*RI. Treat the cut plasmid with phosphatase. Ligate the gene A fragment into plasmid B. Transform.

   c. Cut gene A out of plasmid A using *Eco*RI (cuts on the 5' side) and *Bam*HI (cuts on the 3' side). Digest plasmid B with *Eco*RI and *Bam*HI. Ligate the gene A fragment into plasmid B. Transform.

   d. PCR gene A. The forward primer has an *Eco*RI site in its tail and the reverse primer has a *Bam*HI site in its tail. Cut the PCR product with *Eco*RI and *Bam*HI. Digest plasmid B with *Eco*RI and *Bam*HI. Treat the cut plasmid with phosphatase. Ligate the gene A fragment into plasmid B. Transform.

5. If you cut a plasmid with a restriction enzyme that cuts once and transform bacteria with the restriction digest, would you expect to get colonies? Explain.

6. Why is a "no insert" ligation always included as a control when setting up a ligation of a DNA fragment into a plasmid?

7. You PCR gene A using primers with *Eco*RI sites in their tails. You cut the PCR product with *Eco*RI and gel-purify the fragment. If you don't add a plasmid fragment, but do set up a ligation:
   a. Will the fragment be circularized in the ligation?

   b. If you transform the ligation, will you get colonies? Explain.

8. Explain how this sequence of events is possible. (There are two correct, fairly straightforward kinds of strategies – try to figure out both. And there could be additional more elaborate strategies).
   a. Use one restriction enzyme to cut out DNA fragment 1.
   b. Use a different restriction enzyme to cut out DNA fragment 2.
   c. Ligate fragment 1 to fragment 2.
   d. The ligated DNA cannot be cut with either of the restriction enzymes used in (a) or (b).

Class Activity 3B: Restriction Enzymes and Cloning

9. This map shows the plasmid containing the gene for His-tagged GFP behind a constitutive promoter.

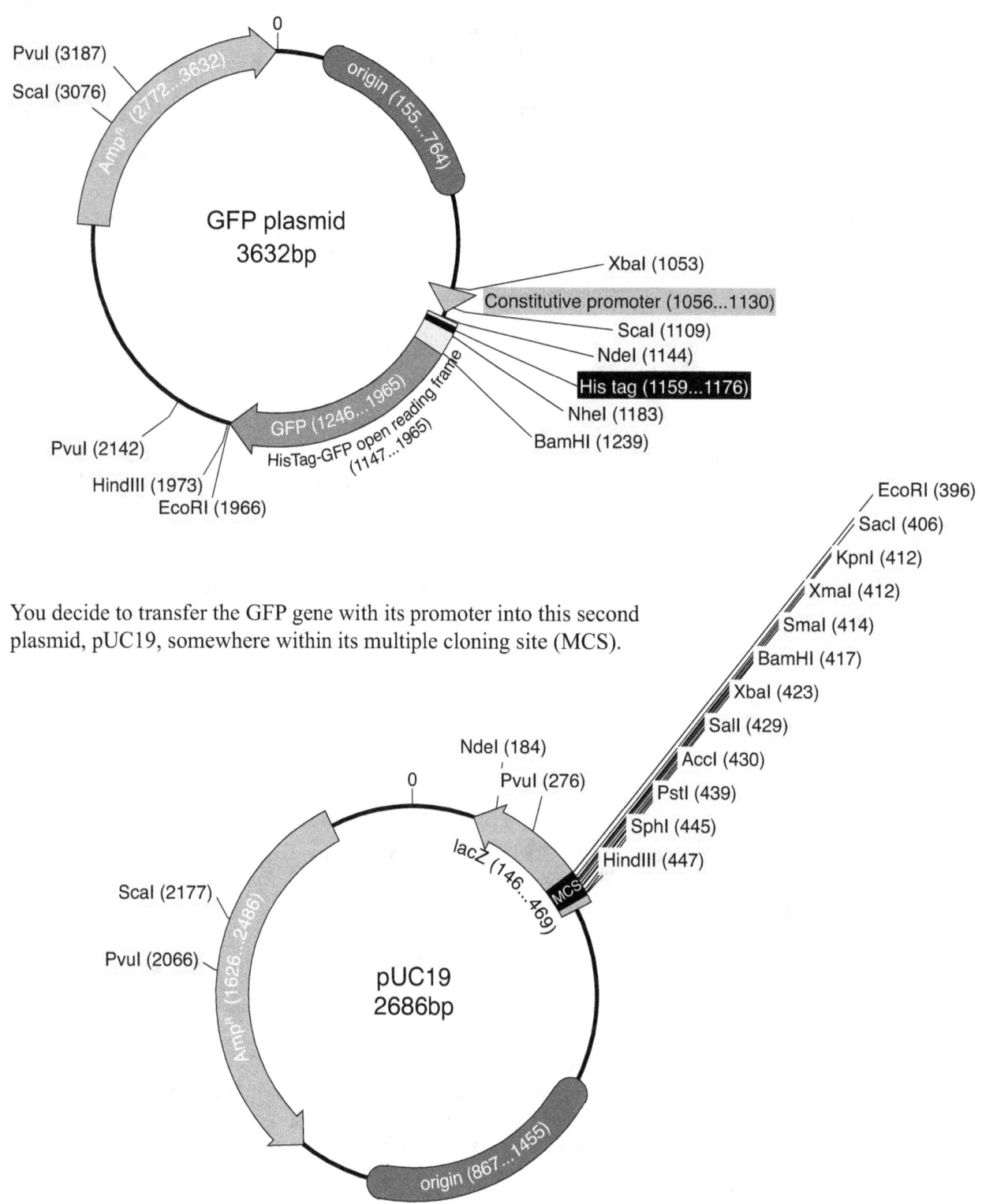

You decide to transfer the GFP gene with its promoter into this second plasmid, pUC19, somewhere within its multiple cloning site (MCS).

a. With which restriction enzyme(s) would you digest each plasmid?

b. What size products would be produced from each digest?

c. Which of these fragments would you ligate together?

60 | Unit 3   *Class Activity 3B: Restriction Enzymes and Cloning*

d. Sketch the plasmid you would make. Include:
   - GFP and its promoter (show direction with arrows)
   - Amp^R gene (show direction with arrow)
   - Origin
   - Locations of the restriction enzyme sites that you used in your cloning.
   - Total size of the new plasmid
   - Base pair numbers at which the restriction enzymes you labeled cut in the new plasmid

e. You've transformed your ligation, have picked several colonies and isolated plasmid DNA from these bacteria. You need to confirm that you've created the plasmid that you wanted to make. You design a restriction digest that will give different fragment sizes if you cut the new plasmid compared to the old plasmids. It's best to choose restriction enzyme(s) that cut at least once in the insert and at least once in the vector so that you create novel fragment sizes.
   i. With which restriction enzyme(s) will you cut?

   ii. Label these restriction enzymes and their base pair locations on the plasmid map you sketched above.

   iii. What size fragments will be produced if you cut:
      The new plasmid?   Original GFP plasmid?   pUC19?

   iv. How will you know if you've made the new plasmid?

   v. You can confirm that you've done everything correctly above (and do this on the computer like you probably would in a lab) by downloading the GFP plasmid and pUC19 sequence files from https://hackettmolecularbiology.blogspot.com/ and using a DNA analysis program to construct the new plasmid and confirm the digests (two solutions are posted). Realize that the maps above and computer programs tend to list the location of the restriction sites based on the beginning of the recognition sequence, so final locations might disagree by a few nucleotides when done on the computer.

Class Activity 3B: Restriction Enzymes and Cloning

Class Activity 3C: Gibson Assembly Mutagenesis of Fluorescent Protein Genes

Work with others in your small class group to complete these questions. Thoroughly discuss one question at a time without skipping ahead. Use your logic skills to answer the questions - do not look up answers in your textbook or in another reference. [The activity below focuses on the mutation of GFP. It's also possible to do this activity to mutate the gene for mCherry or mTangerine to create a different "fruit" color. These mutations are more complex, requiring multiple mutations at once. If you'd like an extra challenge, resources to analyze fruit mutations are also available on https://hackettmolecularbiology.blogspot.com/ in the resources for the wet lab experiment. If you plan to actually do the mutagenesis, note that the rate of success is much lower for "fruit" mutagenesis due to the larger number of nucleotides that need to be mutated compared to GFP mutagenesis].

1. You decide to mutate the gene for green fluorescent protein (GFP) to create a different color protein. (Activity 2B describes fluorescent proteins). Amino acids 66, 67, and 68 of GFP form the chromophore, which absorbs blue light and then emits green light. The following mutations of GFP are known to create different colors:

Mutation	New mutant protein
Y67H	Blue fluorescent protein (BFP)
Y67W	Cyan fluorescent protein (CFP)
T204Y	Yellow fluorescent protein (YFP)

In the table above, note the notation "Y67H" means that the GFP amino acid at position 67 is a tyrosine (Y) and if it's mutated to histidine (H), the new gene encodes BFP. Which color protein will you make? Which amino acid will you mutate and what will the new amino acid be?

2. The sequence below shows the amino acids in the GFP protein.

```
  1 MVSKGEELFT GVVPILVELD GDVNGHKFSV RGEGEGDATN GKLTLKFICT TGKLPVPWPT LVTTLTYGVQ 70
 71 CFARYPDHMK QHDFFKSAMP EGYVQERTIF FKDDGTYKTR AEVKFEGDTL VNRIELKGID FKEDGNILGH 140
141 KLEYNYNSHK VYITADKQKN GIKVNFKIRH NVEDGSVQLA DHYQQNTPIG DGPVLLPDNH YLSTQSVLSK 210
211 DPNEKRDHMV LLEFVTAAGI TLGMDELYK  239
```

The sequence below provides the DNA coding sequence of the GFP gene (as sequenced for the Tsien lab BioBridge GFP plasmid by J. Hackett's students) along with its translation. Mark the position of the amino acid you will mutate in the sequence below. Mark the codon that needs to be mutated.

```
      M   V   S   K   G   E   E   L   F   T   G   V   V   P   I   L   V   E   L   D
  1 ATG GTG AGC AAG GGC GAG GAG CTG TTC ACC GGG GTG GTG CCC ATC CTG GTC GAG CTG GAC  60
      G   D   V   N   G   H   K   F   S   V   R   G   E   G   E   G   D   A   T   N
 61 GGC GAC GTA AAC GGC CAC AAG TTC AGC GTG AGG GGC GAA GGC GAG GGC GAT GCC ACC AAC 120
      G   K   L   T   L   K   F   I   C   T   T   G   K   L   P   V   P   W   P   T
121 GGC AAG CTG ACC CTG AAG TTC ATC TGC ACC ACC GGC AAG CTG CCC GTG CCC TGG CCC ACC 180
      L   V   T   T   L   T   Y   G   V   Q   C   F   A   R   Y   P   D   H   M   K
181 CTC GTG ACC ACC TTG ACC TAC GGC GTG CAG TGC TTC GCC CGC TAC CCC GAC CAC ATG AAG 240
      Q   H   D   F   F   K   S   A   M   P   E   G   Y   V   Q   E   R   T   I   F
241 CAG CAC GAC TTC TTC AAG TCC GCC ATG CCC GAA GGC TAC GTC CAG GAG CGC ACC ATC TTC 300
```

```
          F   K   D   D   G   T   Y   K   T   R   A   E   V   K   F   E   G   D   T   L
301  TTC AAG GAC GAC GGC ACC TAC AAG ACC CGC GCC GAG GTG AAG TTC GAG GGC GAC ACC CTG  360
          V   N   R   I   E   L   K   G   I   D   F   K   E   D   G   N   I   L   G   H
361  GTG AAC CGC ATC GAG CTG AAG GGC ATC GAC TTC AAG GAG GAC GGC AAC ATC CTG GGG CAC  420
          K   L   E   Y   N   Y   N   S   H   K   V   Y   I   T   A   D   K   Q   K   N
421  AAG CTG GAG TAC AAC TAC AAC AGC CAC AAG GTC TAT ATC ACC GCC GAC AAG CAG AAG AAC  480
          G   I   K   V   N   F   K   I   R   H   N   V   E   D   G   S   V   Q   L   A
481  GGC ATC AAG GTG AAC TTC AAG ATC CGC CAC AAC GTG GAG GAC GGC AGC GTG CAG CTC GCC  540
          D   H   Y   Q   Q   N   T   P   I   G   D   G   P   V   L   L   P   D   N   H
541  GAC CAC TAC CAG CAG AAC ACC CCC ATC GGC GAC GGC CCC GTG CTG CTG CCC GAC AAC CAC  600
          Y   L   S   T   Q   S   V   L   S   K   D   P   N   E   K   R   D   H   M   V
601  TAC CTG AGC ACC CAG TCC GTG CTG AGC AAA GAC CCC AAC GAG AAG CGC GAT CAC ATG GTC  660
          L   L   E   F   V   T   A   A   G   I   T   L   G   M   D   E   L   Y   K   *
661  CTG CTG GAG TTC GTG ACC GCC GCC GGG ATC ACT CTC GGC ATG GAC GAG CTG TAC AAG TAA  720
```

3. Determine the simplest mutation you can make to create the desired new amino acid. The genetic code is included in the reference appendix of this book.

You will mutate the GFP gene using a variation of Gibson assembly to reconstruct a mutated GFP expression plasmid. In Gibson assembly, multiple overlapping DNA fragments can be combined into one plasmid all at once in one tube at one temperature by combining the overlapping DNA fragments with three enzymes: a 5' to 3' exonuclease, DNA polymerase, and DNA ligase. The exonuclease chews back all of the 5' ends of the DNA fragments, exposing 3' single-stranded DNA regions. Because the DNA fragments have overlapping sequence, the 3' ssDNA sequences on overlapping fragments will be complementary and will anneal. DNA polymerase fills in any extra gaps left by the exonucleases and DNA ligase seals the remaining nicks in the DNA backbone. This process is drawn in detail for the DNA you will use in a figure later in this activity. Reference: Gibson, D.G., et al. Enzymatic assembly of DNA molecules up to several hundred kilobases. *Nature Methods* **6** (2009) 343-45.

Fragments for Gibson Assembly can be generated either through restriction digest or by PCR. Mutations can be included in the tails of PCR primers used to generate the fragments. Reference: Mitchell, L.A. et al., Multichange Isothermal Mutagenesis: a new strategy for multiple site-directed mutations in plasmid DNA. *ACS Synthetic Biology* **2** (2013) 473-77.

The star in this diagram indicates a mutation introduced in a PCR primer used in a reaction to create one of the DNA fragments for Gibson assembly:

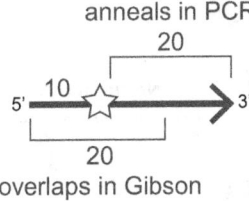

Notice that the mutation cannot be included in the 20nt portion of the primer that anneals to the template in the PCR reaction. However, the mutation can be positioned at the base of the primer's 5' tail. PCR primers used to create adjacent fragments can be used to introduce the new mutation into the overlapping region of both fragments. Because the mutation is part of the overlap, it will be included in the new plasmid following Gibson assembly. About 20nt of overlap are necessary for Gibson assembly.

The plasmid you will mutagenize is depicted below:

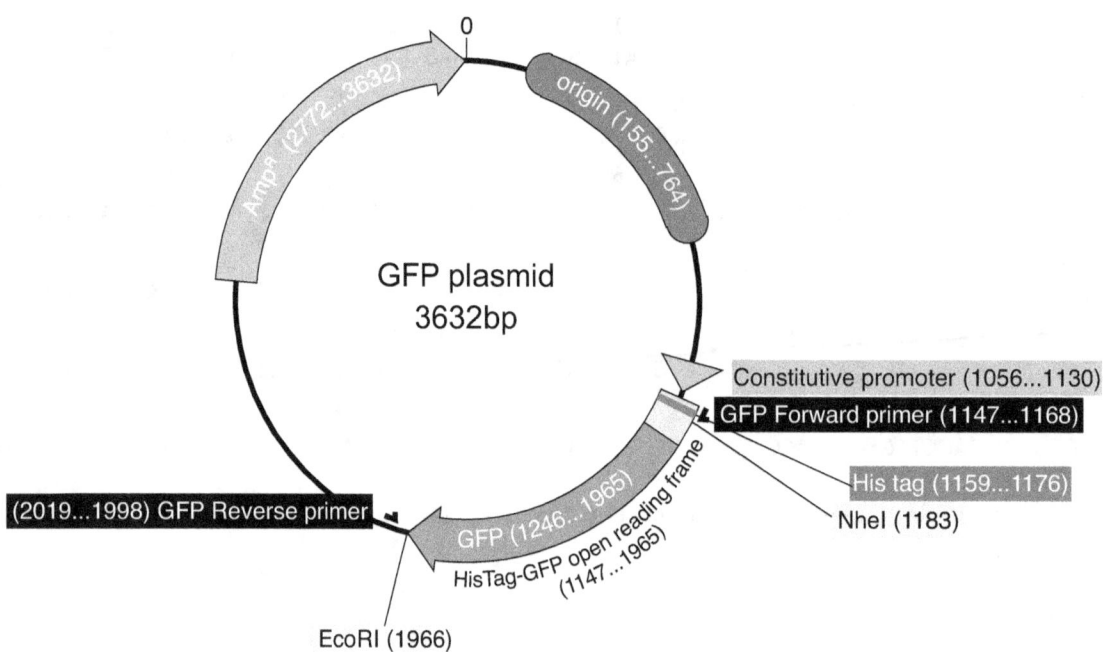

4. One of the DNA fragments you will make is the *NheI/Eco*RI restriction fragment that contains the plasmid backbone, including the ampicillin resistance gene and origin of replication. If you digest the original plasmid with these two enzymes, what size fragments will you produce? Which of these fragments should you save for the Gibson assembly?

You will create two more DNA fragments as PCR products that cover the 5' and 3' portions of GFP and that overlap with each other and with the plasmid backbone restriction fragment. You'll include your mutation (star) in the primers used for the overlap of the two PCR products. This strategy is shown below:

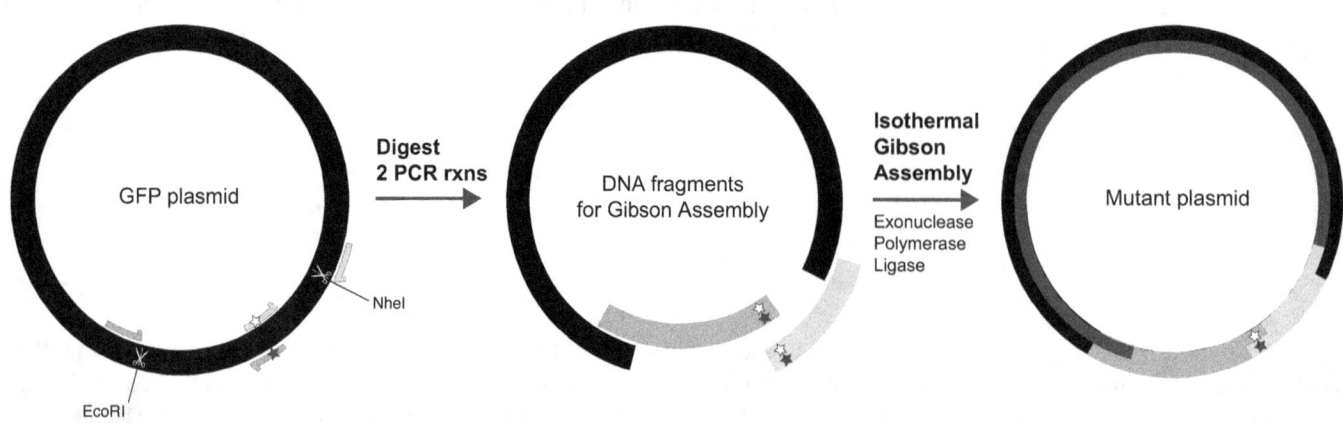

64 | Unit 3   *Class Activity 3C: Gibson Assembly Mutagenesis of Fluorescent Protein Genes*

Your instructor has designed the GFP forward primer and the GFP reverse primer, which are the light gray and dark gray primers that overlap with the plasmid backbone fragment. Your task will be to design the two primers that introduce the mutation, which are the light gray and dark gray primers containing stars in the diagram.

The following diagram shows in detail how the PCR primers should be positioned relative to the mutation:

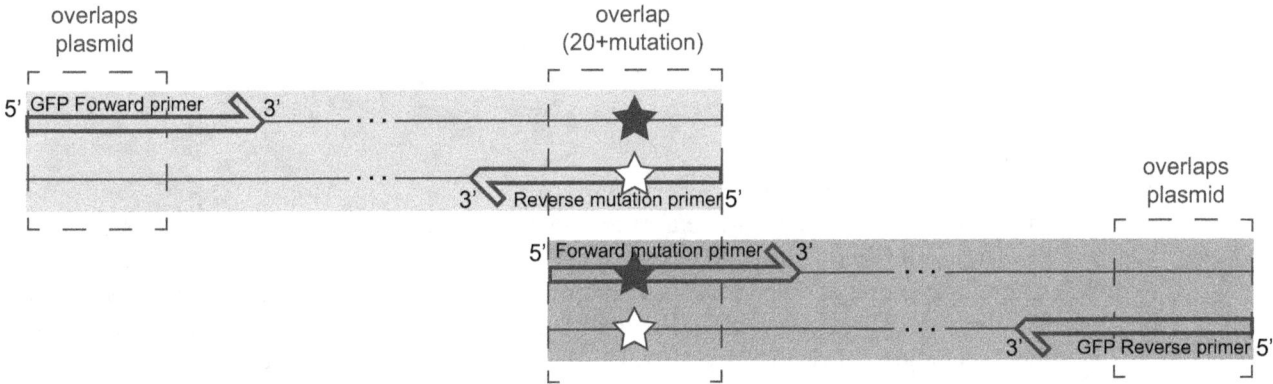

To help you envision how the DNA fragments will be altered and combined by the exonuclease, polymerase and ligase in the Gibson assembly, a detailed version of this process is sketched below. The extreme left and right sides are areas where the two PCR fragments overlap with the plasmid backbone (black), but only a small portion of either side of the plasmid backbone is drawn.

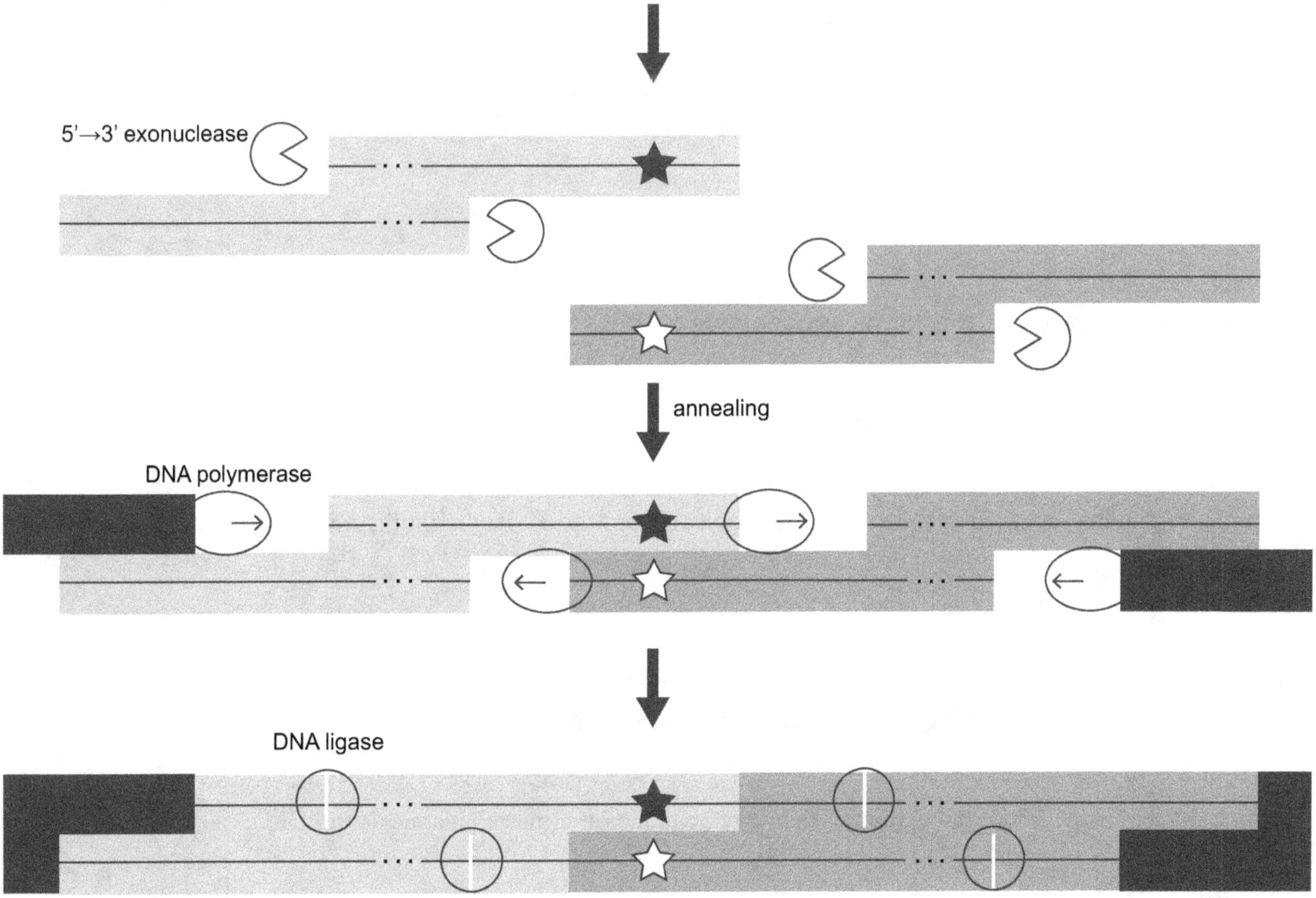

Class Activity 3C: Gibson Assembly Mutagenesis of Fluorescent Protein Genes

Design the sequence of each of the two "mutation" primers, the Forward mutation primer and the Reverse mutation primer. Be sure to write both primers 5' to 3' because that is how primer sequences are accepted by DNA synthesis companies that make primers. Answers can be recorded in the questions below.

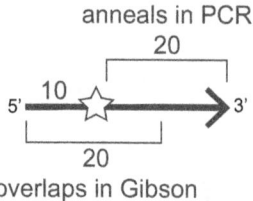

Rules for mutagenesis primer design:
1. Primers for mutagenesis should be about 30bp with a 10bp 5' tail of extra overlap sequence, then the mutation, and then 20bp of 3' sequence that anneals in a PCR reaction.
2. Design primers so that there is at least 20bp of overlap of adjacent DNA fragments that will be joined through Gibson assembly. Usually 10bp of extra overlap is included in the 5'tail of each primer. (For the primers that overlap with the plasmid backbone fragment, the whole 20bp of overlap is included in the one primer).
3. For a single PCR reaction, the melting temperatures (Tm's) of the portions of the primers that anneal to the DNA in the PCR reaction should be similar. A simple way to calculate Tm is to add 4°C for each G/C and 2°C for each A/T.
4. Your instructor has already designed primers that overlap the *Nhe*I side of the cut plasmid backbone (GFP forward primer) and the *Eco*RI side of the cut plasmid backbone (GFP reverse primer). If these two primers are used together in a PCR reaction, they will PCR the whole coding sequence of GFP, and this whole GFP fragment should be used in a control Gibson Assembly reaction. However, you will also be using each of these primers in a separate PCR reaction along with one of your mutation primers to generate the two half-GFP fragments. The Tms of the GFP forward primer and the GFP reverse primer are 60°C.

5. Answer the following questions and check your primer design with your instructor:
   a. Write the sequence of the Forward mutation primer 5' to 3'. Highlight the mutation.

   b. Which primer will this one be paired with in the PCR reaction?

   c. What is the Tm of the portion of the Forward mutation primer that anneals in PCR?

   d. Write the sequence of the Reverse mutation primer 5' to 3'. Highlight the mutation.

   e. Which primer will this one be paired with in the PCR reaction?

   f. What is the Tm of the portion of the Reverse mutation primer that anneals in PCR?

The general steps in Gibson assembly are:
   a. Create DNA fragments that overlap by 20 bp using PCR and/or restriction digest. [Also create fragments for a positive control for the Gibson Assembly].
   b. Remove all of the original plasmid from the solutions of each fragment.
      i. For fragments obtained by restriction digest, be sure the restriction digest has sufficient time to go to completion (the original plasmid should all be digested) and then gel purify the desired fragment – identify it based on its size.
      ii. For fragments obtained by PCR, either of these strategies can be used to remove the original plasmid template:
         1. Gel purify the PCR product
         2. Digest the PCR reaction with *DpnI*. *DpnI* is a restriction enzyme that cuts methylated DNA at the site GATC, but does not cut unmethylated DNA. A methyl group is a -CH_3 group that is covalently linked to a molecule. *E. coli* methylate the GATC site, so the original plasmid which was isolated from bacteria will have methylated DNA.
   c. Gibson assembly: mix overlapping DNA fragments, 5' to 3' exonuclease, DNA polymerase, and DNA ligase and incubate at 50°C for 15-60min.
   d. Transform the assembled Gibson reaction mixture into competent bacteria.
   e. Confirm assembly in individual colonies.

6. Why is it necessary to do the gel purification or *DpnI* digestion of the PCR products?

7. How does the use of *DpnI* allow for the digestion of the original plasmid, but not the PCR product?

8. On average, how often would you predict that *DpnI* should cut methylated DNA? (Once every __ bp). Given the size of the original plasmid that is indicated in the diagram above, about how many times would you predict it should be cut by *DpnI*?

9. Describe how/why Gibson assembly would fail if you left each of the following out of the reaction mixture.
   a. Exonuclease

   b. DNA polymerase

   c. DNA ligase

10. A student uses the Forward mutation primer and the Reverse mutation primer together in the same PCR reaction with the original GFP plasmid as a template. Sketch or describe the PCR product that would be predicted in this reaction. Could this PCR product be used in the Gibson assembly?

Class Activity 3C: Gibson Assembly Mutagenesis of Fluorescent Protein Genes

11. When attempting to mutagenize *GFP* to make *YFP* using the original plasmid described above, you unexpectedly get all white colonies on your ampicillin transformation plate. As a control, you also transform some of the original *GFP* plasmid and get only green colonies. Assuming you initially designed your primers correctly, provide one other logical explanation for what could have happened to produce white colonies in the mutagenesis.

12. Explain the major advantage of Gibson assembly compared to restriction enzyme-based plasmid assembly.

13. Which of the following sets of "mutation" primers are correctly designed (for mutagenesis through Gibson assembly) so that they will introduce a mutation (uppercase nucleotide) between two DNA fragments and have annealing temperatures of 60-62°C with the template DNA using the formula for Tm described above? Explain why you made this choice and rejected the other options.

    a.   5' tccgccatcGcgattgcagatcgataaatgc 3'
           5' aggcggtagCgctaacgtctagctatttacg 3'

    b.   5' tccgccatcGcgattgcagatcgataaatgc 3'
           5' agacgttagcGctaccgcctacatatacggc 3'

    c.   5' tccgccatcGcgattgcagatcgataaatgc 3'
           5' tctgcaatcgCgatggcggatgtatatgccg 3'

    d.   5' tccgccatcGcgattgcagatcgataaatg 3'
           5' aggcggtagCgctaacgtctagctatttac 3'

    e.   5' tccgccatcGcgattgcagatcgataaatg 3'
           5' agacgttagcGctaccgcctacatatacgg 3'

    f.   5' tccgccatcGcgattgcagatcgataaatg 3'
           5' tctgcaatcgCgatggcggatgtatatgcc 3'

Unit 3 Self-Assessment

1. Given the following:

   Interphase   Anaphase
   Prophase   Telophase
   Metaphase   Cytokinesis

   Indicate the FIRST stage of the cell cycle which would not progress properly if:
   a. A defect in a kinetochore protein prevented it from binding tubulin.

   b. The drug Taxol was present. It prevents microtubule disassembly.

   c. There was no primase.

   d. The protease that degrades cohesin is absent.

2. List three features that a small DNA molecule must have in order to be replicated continuously within bacteria. For each, state why this feature is necessary.

3. Genomic sea urchin DNA completely denatures at a lower temperature than genomic turtle DNA. What can you conclude about the relative nucleotide percentages in these organisms? Explain.

4. Sketch the predominant PCR product that will result after ~30 PCR cycles if the two primers (arrows) shown below are used with the plasmid template shown. Show both strands. Include the locations of lettered portions in your product. Also indicate the exact location of the primers (if any) in the product using arrows and numbers - be specific in the strand(s) you indicate.

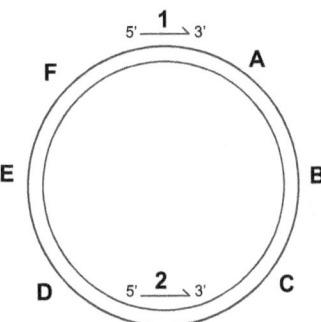

5. Explain how the functions of the following enzymes are replaced in a PCR reaction:
   a. Primase

   b. Helicase

6. Explain the spontaneity (ΔG) of primer annealing in terms of ΔH, ΔS, and T.

7. Compare and contrast PCR and DNA sequencing in terms of:
   a. The components of the reaction.

   b. The products that are produced.

8. Why would it be a problem in cells if Okazaki fragments were not ligated? Give one specific explanation.

9. A particular Taq polymerase inserts the wrong nucleotide once every 1000 nucleotides. However, this Taq polymerase also has 3' to 5' proofreading activity that allows it to replace the wrong nucleotide most of the time. If the overall mutation rate is 1 mutation out of 100,000 nucleotides (10^{-5}), what percentage of the time does proofreading happen correctly? Show your work.

10. Complete the following table with the volumes of each reagent needed to set up this PCR reaction.

Component	Stock Concentration	Final concentration, mass, or units	Volume
Template DNA	50 ng/μL	2 ng/μL	
Primer 1	4 μM	0.2 μM	
Primer 2	4 μM	0.2 μM	
Taq/buffer/dNTP master mix	2X	1X	
Water	n/a	n/a	
		Total Volume:	50 μL

11. a. Explain why STR analysis (PCR) has replaced the restriction enzyme digest/Southern blotting method of DNA fingerprinting.

    b. Why is Southern blotting preferable to PCR for confirming the deletion of a gene in the genome (in genetic engineering)?

12. State two examples of proteins that interact with DNA in a manner that is **not** sequence-specific.

13. Give one example of a protein that interacts with DNA in a manner that is sequence-specific.

14. The following questions refer to this DNA gel in which a student is analyzing different food samples using PCR to determine whether they have been genetically modified. The "tub" PCR copies a section of the tubulin gene which is found in all plants. The "35S" PCR copies a section of the plasmid used to genetically modify plants.

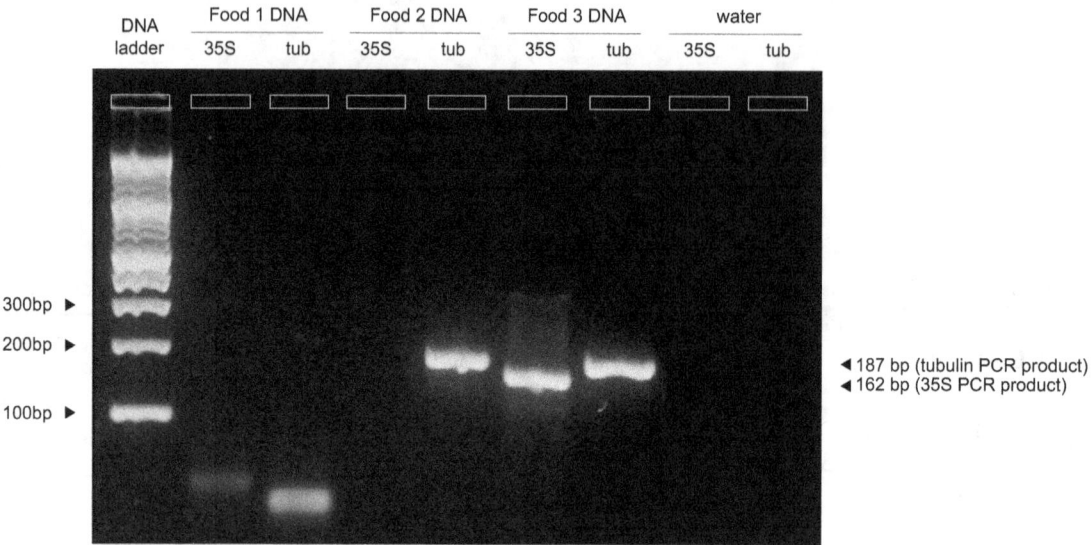

   a. Label the image with "+" and "–" to indicate the relative locations of the electrodes.
   b. Write the conclusion you would draw based on these results for each of the food samples:
      i. Food 1

      ii. Food 2

      iii. Food 3

15. A scientist decides to delete the yeast gene *LIG4* in the yeast genome and replace it with the gene *KanMX* that allows the yeast to grow on the antibiotic G418. The yeast are haploid.
   a. Describe the structure of the DNA molecule that will be transformed into yeast in order to accomplish this. Through what process will the replacement take place?

   b. *LIG1* encodes the DNA ligase used in DNA replication. *LIG4* encodes the DNA ligase used in the repair of double-strand breaks. What specific process would be defective in the yeast made above (*lig4*-deleted)? How would you predict the mutation rate in the *lig4*-deleted yeast would compare to the wildtype mutation rate? Explain.

16. Explain why it makes sense that many restriction enzymes are dimers.

17. Write a 6-nucleotide sequence that is a likely recognition site for a restriction enzyme that is a dimer.

18. *Sma*I is a blunt-cutting restriction enzyme. You decide to clone a PCR product into the *Sma*I site of a plasmid. Prior to ligation and transformation:
    a. How will you have to treat the PCR product?

    b. How will you have to treat the plasmid?

19. What function do restriction enzymes have in bacteria?

20. Explain why the bacterial genome isn't digested by restriction enzymes.

21. You decide to clone Gene E into a plasmid and express it in bacteria. Gene E is a human gene that is expressed in white blood cells. Starting with white blood cells, bacteria, and any genetic engineering supplies, outline the steps you would need to take in order to obtain a plasmid that is ready to be transformed into bacteria and that will express gene E. Note: one of your "genetic engineering supplies" is an Amp^R plasmid that contains a constitutive promoter and RBS followed by an *Eco*RI restriction enzyme site and a *Bam*HI restriction enzyme site.

22. In a Gibson mutagenesis experiment, assume that a student's design was to make a single base substitution that changed green fluorescent protein (GFP) into blue fluorescent protein (BFP). When they plated the bacteria from their transformation with the Gibson assembly reactions, they got some blue colonies and some colonies of other colors on the same ampicillin plate (under UV light). They used the GFP plasmid as a template for their PCR reactions and used Gibson Assembly to insert their PCR products into a plasmid backbone from which GFP had been cut out.
    a. Explain what might have happened in the experiment to give them:
        i. A blue colony

        ii. A green colony

        iii. A white colony

    b. In this experiment described in the previous question, what will the student need to do in order to confirm that they've actually mutagenized the plasmids as they've planned to do? (Simply qualitatively observing a color isn't enough - what else needs to be done)?

23. The restriction enzyme *Not*I has an 8 bp recognition sequence. How many times would you predict *Not*I would cut a 100 million bp human chromosome?

24. In the following plasmids, *Xba*I, *Nde*I, *Bam*HI, *Eco*RI, and *Hind*III are restriction enzyme cut sites. Each restriction enzyme produces unique sticky ends. Numbers indicate the positions of the cut sites. The total plasmid size is indicated in the center. Promoters (constitutive in #1 and inducible P$_{BAD}$ promoter in #2) are indicated by bold arrows that indicate the direction of transcription from that promoter. You are asked to move the GFP coding-sequence from plasmid #1 (where it is currently expressed from a constitutive promoter) into plasmid #2 so that it can be expressed from the P$_{BAD}$ promoter. The P$_{BAD}$ promoter is only "on" in the presence of the sugar arabinose.

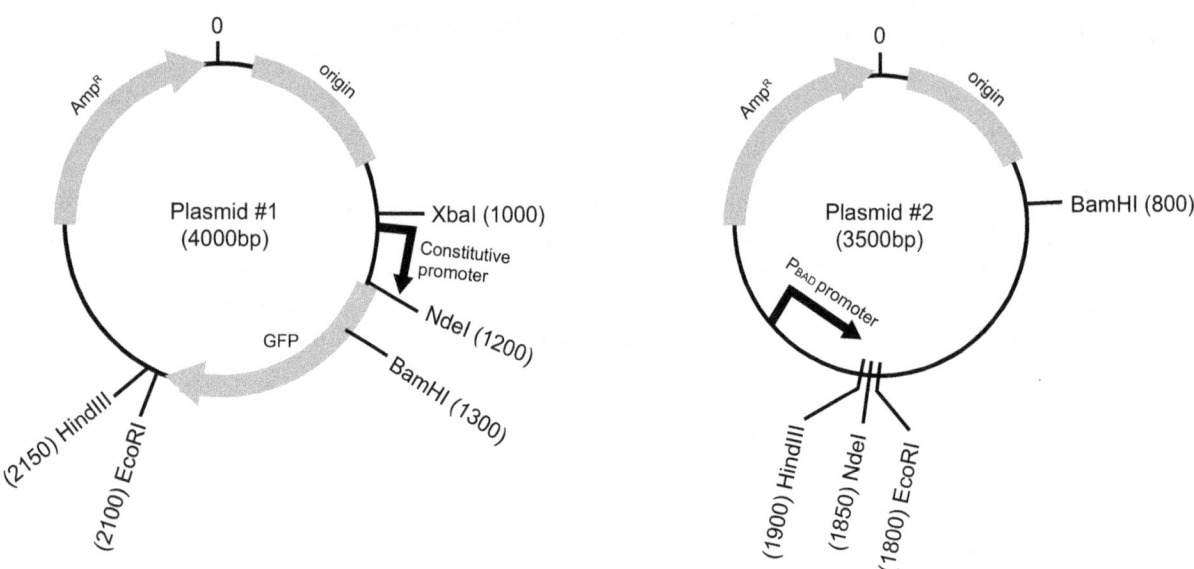

a. Sketch the new plasmid you will make, including the positions of all restriction enzyme sites.

b. If you get colonies on your transformation plate, what would you need to do to confirm they were "correct"?

c. What size fragment(s) will be produced if your new plasmid is digested with *Bam*HI?

Unit 4

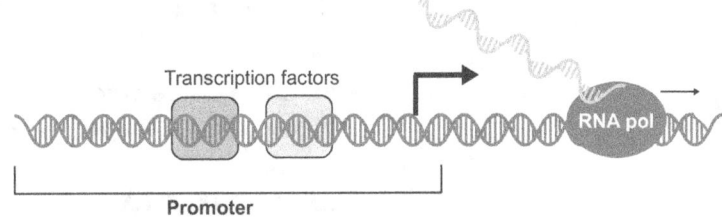

The Regulation of Gene Expression

A person's genotype (genetic makeup) is created by the sequence of the person's DNA. A person's phenotype (observable physical/biochemical characteristics) is determined by the combination of the person's genotype, how the person's genes are expressed and the impact of environmental factors. This unit will introduce how specific DNA sequences and modifications to DNA can regulate transcription. In addition, cells possess additional mechanisms for regulating the abundance of mRNA molecules following transcription, thus regulating the level of translation. While an organism's genome contains the genes that code for the proteins needed by that organism, it is the regulation of the expression of those genes that allows cells to respond to environmental stimuli and that creates the variety of cell types in a multicellular organism.

To achieve the deepest level of understanding, you are encouraged to complete the class activities, textbook readings and question sets in the order listed below.

Suggested order:
   Chapter 4 Reading #1 and Discussion Questions
   Project 4A: Gene Parts - Analysis of an iGEM Project
   Class Activity 4B: Golden Gate Assembly of Yeast Gene Parts
   [Experiment: Golden Gate Assembly of Yeast Gene Parts]
   Chapter 4 Reading #2 and Discussion Questions
   Chapter 4 Reading #3 and Discussion Questions
   [Experiment: Induction of RNAi]
   Chapter 4 Reading #4 and Discussion Questions
   Chapter 4 Reading #5 and Discussion Questions
   Class Activity 4C: Epigenetic Inheritance
   Chapter 4 Reading #6 and Discussion Questions
   Unit 4 Self-Assessment Questions

Prior knowledge (review if necessary):
- Structure of DNA, RNA, and proteins (in Chapters 1, 2, 3)
- Complementary DNA and RNA nucleotides (in Chapters 1, 3)
- How DNA codes for RNA which codes for protein (in Chapter 2)
- mRNA structure (in Chapter 2)
- mRNA processing in eukaryotes (in Chapter 2)
- Factors influencing the binding of proteins to DNA and other proteins (in Chapters 2 and 3)
- Enzyme structure and function (in Chapter 2)
- Organization of genes on chromosomes (in Chapter 3)
- Plasmid construction through genetic engineering (in Chapter 3)
- cDNA synthesis from mRNA (in Chapter 3)

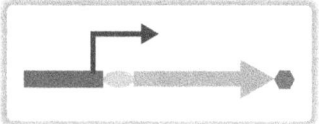

Chapter 4 Reading #1 and Discussion Questions

Reading: *Molecular Biology: Concepts for Inquiry*
Chapter 4: Introduction and Section 4.1A

Online Quiz: 4.1A

Discussion Questions:

1. What is meant by a gene being "on"? What is meant by "gene expression"?

2. What is the role of gene expression in determining the phenotype of an organism? What else contributes to the phenotype?

3. What are the parts of a gene? Why do genes require more parts than just a protein-coding sequence?

4. Why might it be advantageous for a gene to have a promoter that is either inducible or repressible rather than having a constitutive promoter? Give an example of a situation in which you would want to use a constitutive promoter in genetic engineering. An inducible promoter?

5. Would you consider the regulation of gene expression by promoters, transcription factors, enhancers, and repressors to be simple or complex? Speculate on the significance of this in terms of the needs of an organism.

6. Transcription factor X can bind the promoter of both gene A and gene B. Gene A is expressed strongly, but gene B is expressed weakly. What might account for this difference?

7. Examine the iGEM BioBricks in Figure 4.6. Describe a plasmid you could create using these BioBricks. Check that you've included all of the necessary parts of a gene. What would your plasmid do?

8. Bacteria containing a plasmid with a *GFP* gene and an ampicillin resistance gene are bright green when grown in media A, but are white when grown in media B. Both types of media contain ampicillin. What would you predict is true about the structure of the *GFP* gene? If you did minipreps to purify plasmid DNA from samples of bacteria in both types of media, what would you predict would be the result in terms of the yield of DNA? Explain.

9. ⚙ iGEM teams sometime use a DNA device that they call an "inverter" for the purpose of inverting the pattern of transcription of a gene. An inverter turns transcription of a particular gene off under the circumstances that it would otherwise be on. An inverter also turns transcription on under the circumstances that it would otherwise be off. An inverter is not a single gene part. Rather, it is a composite of multiple gene parts (and those parts are the kinds described in the reading). Propose what kinds of parts would be needed to create an inverter and how they would be assembled.

Project 4A: Gene Parts - Analysis of an iGEM project

This project will challenge you to apply your understanding of genes, RNA, proteins, and the regulation of transcription and translation to figure out an iGEM project. There is an additional challenge inherent in this project since the main source of your information will be a wiki page written by undergraduate students – and the clarity of the information on those pages can be variable. When choosing a project to work on, in addition to finding one that's interesting, be sure to find one where the team actually built a full device and entered it into the iGEM parts registry (as opposed to just planned to do something or just built separate parts) and also wrote a clear wiki page.

1. Review the iGEM 3A DNA Assembly method: http://parts.igem.org/Help:Assembly/3A_Assembly

2. Choose one of the synthetic biology projects presented at one of the iGEM Jamborees. Choose a project that you think has an application that has some potential to be beneficial to society.

    a. iGEM team projects are listed on this site: http://igem.org/Previous_iGEM_Competitions

    b. Click on the "Team Abstracts" link for any year to see summaries of the projects. Find out more about a particular project by clicking on a link to go to that team's wiki page. Also click on the "Results" link to see which ones won awards and so are considered the "best" by others. An award for best wiki indicates that their website may be more understandable, but it doesn't necessarily mean that their science is any good. Choose a site with good science (device parts in the registry - see below) AND with an understandable wiki.

    c. Be sure to choose a project in which the team created a new "device" (that consists of a full gene -multiple parts- assembled in a plasmid) or a new "system" (a group of interacting devices assembled in a plasmid or plasmids), rather than a project in which a team built a new single part (like just a promoter), but didn't assemble multiple parts into devices. Definitions for parts and devices are available here: http://parts.igem.org/Help:Parts. Be sure that they actually built the whole plasmid(s) by finding their BioBrick part numbers (usually in a submitted parts page on their wiki) and finding that device "part" in the parts registry (usually by clicking on the BioBrick part number) or go to http://parts.igem.org/Main_Page and type in the number. It's not enough for them to have included cartoon drawings of what they were intending to build in their wiki, they need to have actually built whole device plasmid(s) and entered them into the registry. In the parts registry, a "device" will have at a minimum, all of the parts of a gene (promoter, RBS, ORF, terminator), each of which can be clicked on separately for more information. Be careful - if they actually didn't build anything, they might still have a really good wiki that includes things like plasmid sketches. If your team used device(s) built by others in a novel way, it's fine to choose their project, but you'll probably need to do research on the original team's wiki too.

    d. Double-check that the project you chose meets all of the requirements above. Otherwise, you're likely to get stuck along the way.

3. After finding an acceptable project, do as much research as you can on your own. Try to answer all of the questions in #4 as best you can - this might be difficult to do at first. You may need to do research beyond the team's wiki page. It's common for the students who create the wiki pages not to explain things that they think scientists should already know about. This creates a challenge for you since you're not a professional scientist (yet). Common things that teams assume you know, but that you'll actually have to research are certain regulatable promoter systems like P_{BAD}/AraC, lac, trp, lux, etc. You can find more information about how many of these work in Appendix 3.

4. After choosing a project, research the project on that team's site (and possibly using outside resources). Create a report about the project that includes the following elements (in your own words using language a classmate could understand):

   a. An explanation of the goal of the project – what did they set out to accomplish and why did they want to accomplish it? Be sure to indicate the team name and year.

   b. A sketch of the BioBrick plasmid(s) used to accomplish this goal. Clearly label each of the BioBrick parts in each plasmid- identify which parts are ORFs, promoters, enhancers, repressor binding sites, RBSs, terminators, etc. Use the parts registry page to figure out the actual order. Draw ORFs and promoters using arrows so that their directions are clear. Sketches on the team's wiki or in the parts registry may be incomplete, but your sketch must be complete. Don't forget to include the drug resistance gene(s) and origin. These parts aren't listed on the parts registry page, but they're actually in the plasmid. You can see an origin and drug resistance gene in this plasmid by clicking on the "view plasmid" link on the parts registry page.
      - Write the BioBrick number of the device plasmid in the center of the circle so it can be easily found in the parts registry upon grading.
      - Label the parts in the device with their actual names, not with the BioBrick numbers, so they can be understood.
      - For parts like ORFs and promoters, specify the name of the ORF or promoter. Don't just label it "ORF" or "promoter."
      - Be sure to indicate the direction of promoters and ORFs.
      - Sketch the functional "device" plasmid(s) that has all of the parts in one plasmid rather than sketching the parts separately.
      - Be sure to include any drug resistance genes and origins.
      - Label every part, including ones that the iGEM team might not have labeled because they weren't BioBricks, like the promoter, RBS, and terminator for a drug resistance gene. Sometimes the iGEM team doesn't label important parts even in the parts registry- they might show an ORF part when they really mean that it's an RBS-ORF-terminator (especially common when all 3 parts were together in a genome somewhere originally and they all happened to be cloned as one piece rather than as separate parts). Use your understanding of gene parts to draw the plasmid correctly showing every part. If there are 2 protein-coding parts right next to each other, is there an RBS between them or are they fused into one ORF? If a protein is tagged with another polypeptide, label the tag too.
      - If there is not extra plasmid sequence between 2 parts, put them right next to each other in your sketch. If there is extra plasmid sequence between 2 parts, don't put them right next to each other - leave an unlabeled part of the plasmid circle between them. No plasmid is only composed of gene parts.
      - If your system has more than one plasmid don't combine them into one plasmid. Sketch the real plasmid(s).
      - If your team worked on multiple projects, you're only required to talk about one of them.
      - If as part of one project, your team made multiple similar versions of a plasmid, you're only required to talk about one of them.
      - If as part of one project, two different "device" plasmids were required to work together to create a "system", you probably need to sketch and describe both.

c. A list of every part in your plasmid(s) and a brief statement of its function. This will serve as a reference key for your plasmid sketch. For example, part of this list might be:
    i. *AraC* promoter - repressible promoter that is off when bound by AraC protein and on in the absence of AraC protein. When on, AraC mRNA is produced.
    ii. RBS (for *AraC* gene) - location on AraC mRNA where a ribosome binds to before starting translation. This BioBrick RBS is a strong RBS.
    iii. *AraC* ORF - protein coding sequence for the AraC transcription factor that binds arabinose. AraC protein always represses the AraC promoter. AraC protein represses the P_{BAD} promoter in the absence of arabinose and AraC induces the P_{BAD} promoter when AraC binds arabinose.
    iv. terminator (for *AraC* gene) - causes RNA polymerase to stop transcribing the *AraC* gene. This is the original *AraC* terminator from the E.coli genome.

If some of your plasmid parts work in prokaryotes and some work in eukaryotes, indicate which parts work in which organisms.

d. A paragraph providing a thorough explanation of how the most significant parts of your plasmid(s) and any other relevant molecules in the cell/environment work together as a "device" or "system" to perform a function. You're most likely to need to focus on promoters and proteins encoded by ORFs in this section as opposed to RBSs and terminators. This section should be a clear story that explains step-by-step how things work in a narrative format, rather than a disjointed list of unconnected ideas. Specify what organism contains the plasmid. Be clear about when you're talking about DNA and when you're talking about a protein. For example, you can't just say that "A binds B." Do you mean protein A binds protein B, or do you mean that protein A binds the promoter of gene B? Also, if a molecule that's not encoded by your plasmid is important to the story, be sure to indicate what it is and where it's coming from. Yes, parts of this paragraph will be redundant with part of (c) in terms of the parts described, but the explanation here will need to be as detailed as you can make it (whereas part c descriptions were brief). Also, here, your explanation flows in a narrative. Do not leave things out because you already said them in c.

e. Write a paragraph evaluating this project. Where do the researchers think they should go from here? Where do you think they should go from here? Do you think that the device they've constructed is potentially useful?

Class Activity 4B: Golden Gate Assembly of Yeast Gene Parts

Work with others in your small class group to complete these questions. Thoroughly discuss one question at a time without skipping ahead. Use your logic skills to answer the questions - do not look up answers in your textbook or in another reference.

It can be useful to genetic engineers to be able to assemble genes from their parts. You explored one method for doing this in the previous iGEM project that used "3A" assembly. In this strategy, parts are assembled based on cutting with two restriction enzymes that cut within their recognition sites and that generate complementary sticky ends in such a way that the junction of two adjacent ligated parts no longer has the recognition sequence of either restriction enzyme.

"Golden Gate" assembly (Engler, C., *et al.* A one pot, one step, precision cloning method with high throughput capability. *PLoS ONE* 3, e3647 *(*2008)) employs an alternate strategy for gene assembly that relies on a different kind of restriction enzyme. *Bsa*I is a type IIS restriction enzyme, meaning that it cuts at a sequence outside of its recognition site. In the diagram below, "N" could be any nucleotide:

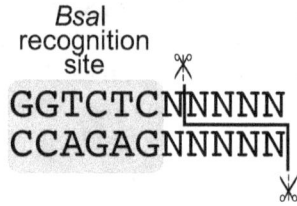

Therefore, *Bsa*I can be used to generate sticky ends of any possible sequence. DNA fragments generated using *Bsa*I can be ligated together if they have complementary sticky ends.

We will study the assembly of the parts of a yeast gene using the Golden Gate method as described in: Boeke, J., *et al.* Yeast Golden Gate: Standardized assembly of *S. cerevisiae* transcriptional units. BBF RFC 88 (2012) available at http://hdl.handle.net/1721.1/73912.

1. Given the above description of how *Bsa*I recognizes and cuts DNA, why does it make sense that its recognition sequence is not a palindrome?

We are going to focus on the assembly of a gene that can express green fluorescent protein (GFP) in yeast under the control of the inducible *GAL1* promoter. Expression from this yeast promoter is repressed when yeast are grown in typical media containing the sugar glucose. Expression from the *GAL1* promoter is induced 1000-fold when the sugar galactose is substituted for glucose in the media. The *GAL1* promoter, the *GFP* protein-coding sequence (ORF, open reading frame), and the *MFA2* terminator that contains a 3'UTR and terminator will be assembled together and ligated into a plasmid that can be grown in either bacteria or yeast. The following figure shows the gene parts and plasmid that will be assembled.

GAL1 promoter reference: Johnston, M. & Davis, R.W. Sequences that regulate the divergent *GAL1-GAL10* promoter in *Saccharomyces cerevisiae*. *Molecular and Cellular Biology* 4, 1440-48 (1984).
MFA2 terminator reference: Yamanishi, M., *et al. TPS1* terminator increases mRNA and protein yield in a *Saccharomyces cerevisiae* expression system. *Biosci. Biotechnol. Biochem.* 75, 2234-36 (2011).

Gene parts:

Each part is located in a bacterial plasmid containing a bacterial orign and a kanamycin resistance (*Kan*^R) gene as a selectable marker. Only the portion of the plasmid between and including the *Bsa*I sites is shown.

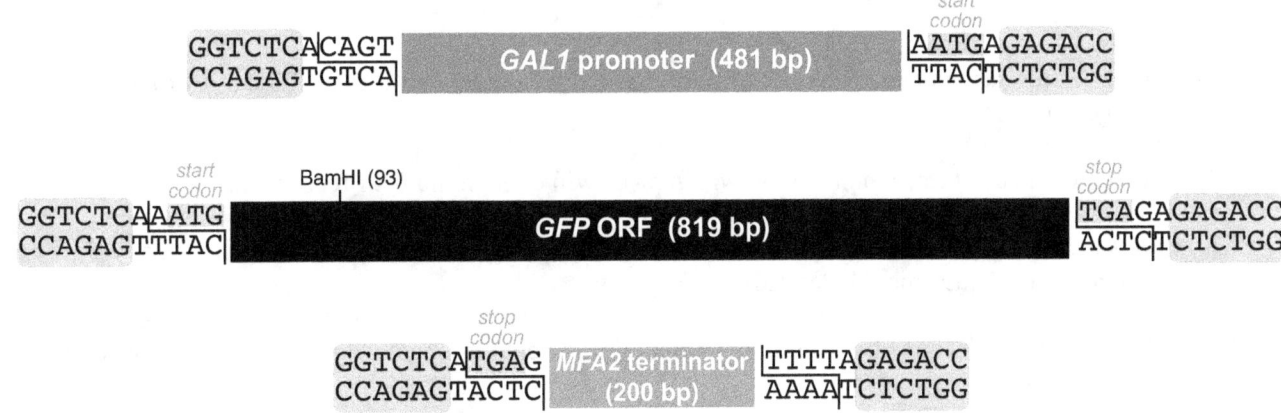

Acceptor vector:

Gene parts will be ligated into this vector. *Amp*^R is the bacterial ampicillin resistance gene. "Origin" is a bacterial origin of replication. *RFP* is a bacterial red fluorescent protein gene. *CEN/ARS* is a yeast centromere and autonomously replicating sequence (origin of replication). *URA3* is a yeast gene required for the synthesis of the nucleotide uracil. Yeast lacking *URA3* will only grow if uracil is added to the growth media. Therefore *URA3* can be used as a selectable marker when yeast are plated on media lacking uracil.

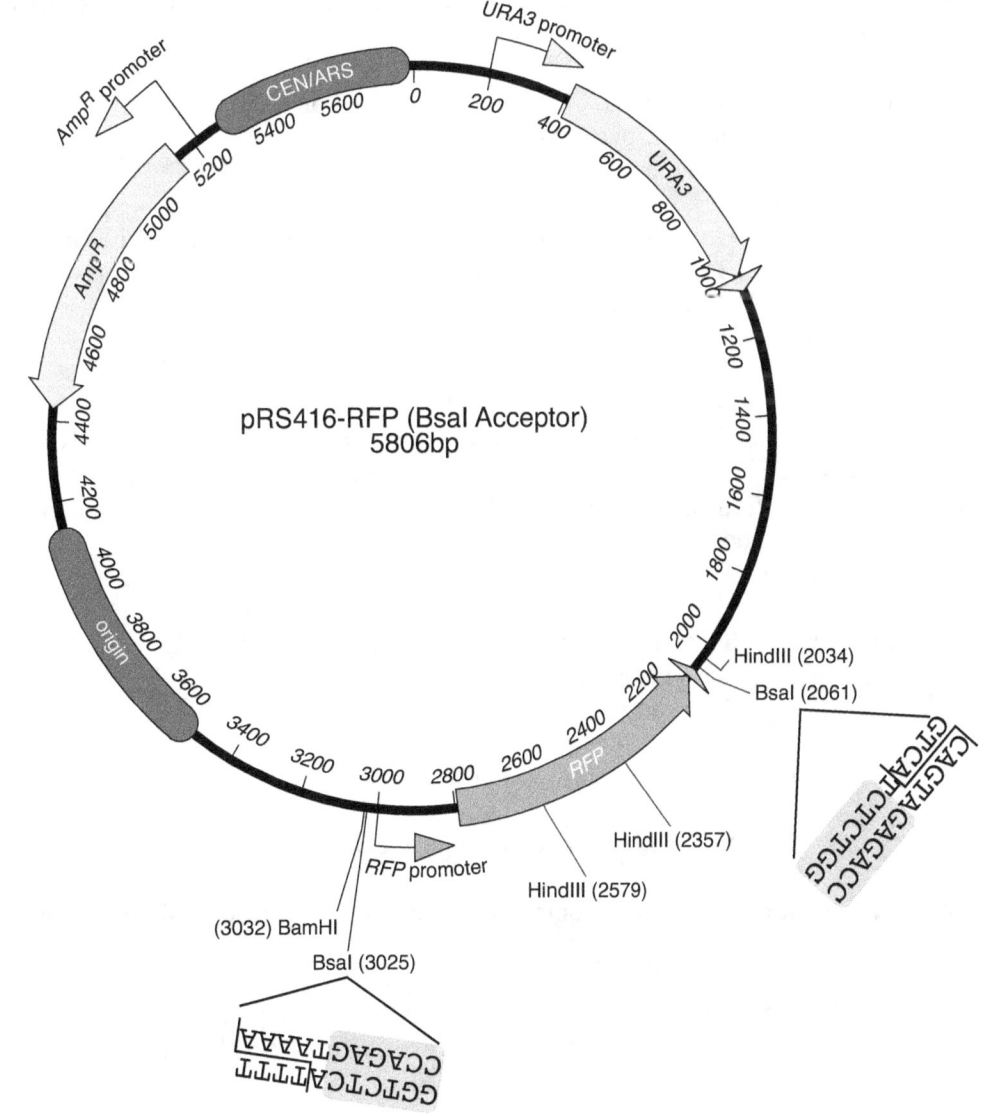

Class Activity 4B: Golden Gate Assembly of Yeast Gene Parts

2. Carefully study the figure above and answer the following:
   a. How do you think the *Bsa*I sequences were introduced when the gene parts were originally cloned into the *Kan*^R gene part plasmids?

   b. When *Bsa*I cuts the gene parts out of their original plasmids, will there still be a *Bsa*I recognition site on the gene parts?

   c. When *Bsa*I cuts the acceptor vector, which piece will contain the *Bsa*I recognition sites?

   d. When the gene parts are ligated to each other and ligated into the acceptor vector, how many *Bsa*I sites will be on the newly assembled plasmid?

   e. Figure out which ends of which *Bsa*I restriction fragments can ligate together. How do you know?

   f. What is the benefit of locating the start codon and stop codon of the protein-coding sequence in the single-stranded overhangs of the ORF *Bsa*I restriction fragment rather than in the double-stranded portion of the gene part?

   g. After the gene parts are ligated into the acceptor vector, will the newly-assembled plasmid contain the *RFP* gene?

   h. What bacterial selectable marker gene is contained on:
      i. The gene part plasmids
      ii. The acceptor vector

3. You combine the following in the same 0.2 mL tube in a buffer solution for the Golden Gate reaction:

       3 gene part plasmids (promoter, ORF, and terminator plasmids)
       Acceptor vector
       *Bsa*I
       DNA ligase

   You place this one tube into a thermal cycler for 60 minutes at 37°C (*Bsa*I and DNA ligase are both active), 5 min at 50°C (*Bsa*I is active, but DNA ligase is not), and 5 min at 80°C (denatures the enzymes). This Golden Gate reaction is then transformed into bacteria and plated onto plates containing an antibiotic.

   a. Which antibiotic, ampicillin or kanamycin, should be included in the plates for the Golden Gate transformation?

   b. If bacteria are transformed with each of these plasmids, will colonies grow on the above plates? Why? If colonies grow, what color will they be (red or white)?
      ii. Gene part plasmid?

      iii. Original acceptor vector?

      iv. Acceptor vector containing gene parts?

   c. What color bacteria should contain the assembled *GFP* gene? Will they express *GFP*? Why or why not?

   d. Can the original acceptor vector be put back together again in its original form after being cut with *Bsa*I during the Golden Gate reaction? What two factors limit the amount of original acceptor vector that is transformed?

4. What was the purpose of including *RFP* in the acceptor vector?

5. Explain how the Golden Gate strategy allows for the assembly of multiple DNA fragments at the same time in one tube. What is clever about this strategy?

6. Using the plasmid maps above and the fragment lengths that are given:
   a. Sketch the new plasmid that you made. Include Amp^R (& promoter), origin, CEN/ARS, *URA3* (& promoter), *GFP*, the *GAL1* promoter, *MFA2* terminator, *Bam*HI sites, and *Hind*III sites. Indicate the total length of your new plasmid and the locations of the *Bam*HI sites and HindIII sites to the nearest base pair. The drawing does not need to be to scale, but the numbers should be correct relative to this new plasmid (not the old one). The orientations of the parts also must be correct and the direction of any part that has a direction (like a promoter or ORF) should be indicated.

   b. You decide to check the plasmid DNA from several colonies from the Golden Gate transformation using a restriction digestion to see if the plasmids seem to be correct prior to sequencing them. To the nearest base pair, list the sizes of the fragments you would expect from:
      i. A *Bam*HI digest of the original *RFP* acceptor vector
      ii. A *Bam*HI digest of the new *GFP* plasmid
      iii. A *Hind*III digest of the original *RFP* acceptor vector
      iv. A *Hind*III digest of the new *GFP* plasmid
      v. A double *Bam*HI/*Hind*III digest of the new *GFP* plasmid

7. You transform your new plasmid into yeast lacking the *URA3* gene and plate them.
   a. What characteristics should the media in the plates have to allow for the growth of only yeast that have the plasmid?

   b. Yeast are normally white. Under what conditions will the yeast be green? Why?

8. Which gene(s) on your new plasmid should/can be expressed in bacteria? Which gene(s) on your new plasmid should/can be expressed in yeast?

9. What factors determine whether a gene can be expressed in a particular species?

10. What are the advantages or disadvantages of using restriction enzyme digests vs. DNA sequencing to verify that a clone is "correct?"

11. If you had a large collection of yeast promoter parts between *Bsa*I sites in Kan^R plasmids and you wanted to find the strongest promoters in the collection, how might you use Golden Gate assembly to help you test the promoters rapidly? How many Golden Gate reaction tubes would you need?

12. What are the similarities and difference between Golden Gate gene assembly and iGEM 3A DNA Assembly of BioBricks: http://parts.igem.org/Help:Assembly/3A_Assembly? Consider the standardization of parts, modularity of parts, scars at the junctions of ligated gene parts, and number of parts that can be assembled at one time. If you were going to establish a new collection of gene parts, which assembly method would you prefer and why?

Chapter 4 Reading #2 and Discussion Questions

Reading: *Molecular Biology: Concepts for Inquiry*
Chapter 4: Section 4.1B-C

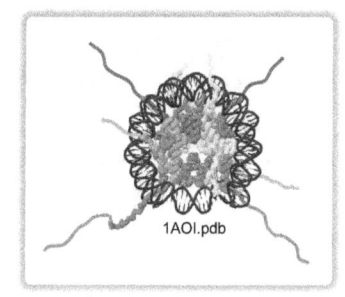

Online Quiz: 4.1B-C

Discussion Questions:

1. What is the advantage of winding DNA around histones? What is the advantage of further winding of DNA? In what form is chromatin found in an interphase cell? Mitotic cell?

2. What role does the formation of either euchromatin or heterochromatin have in the regulation of gene transcription?

3. How does DNA methylation affect the modification of histone tails? What role does methylation play, both directly and indirectly, in the regulation of transcription?

4. What evidence supports the hypothesis that the organization of chromatin into topologically associated domains might contribute to gene expression pattterns?

5. How is the DNA methylation pattern maintained following DNA replication?

6. You decide to insert a new gene into the human genome. Explain how the location at which the gene is inserted into the genome might affect its transcription.

7. Inactivation of DNA methyltransferase through mutation is sometimes observed in cancer cells. How do you think these mutations might be advantageous for the tumor?

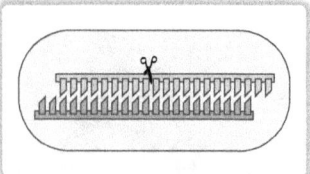

Chapter 4 Reading #3 and Discussion Questions

Reading: *Molecular Biology: Concepts for Inquiry*
           Chapter 4: Section 4.2A-B

Online Quiz: 4.2A-B

Discussion Questions:

1. Explain how the characteristics of eukaryotic mRNA differ from prokaryotic mRNA in ways that allow the eukaryotic mRNA to form looped structures. What is the advantage of forming these looped mRNA structures?

2. What was the conclusion of Fire and Mello's RNAi experiment? What evidence supports this conclusion? Why do you think Fire and Mello decided to include the GFP RNAi experiments in their study instead of only studying RNAi against *unc-22*?

3. How does RNAi explain the puzzling experiments in the petunia, *N. crassa*, and *C. elegans* that preceded the discovery of RNAi?

4. The enzymes involved in RNAi, Dicer and Argonaute, are not capable of recognizing specific RNA sequences. Explain how the action of each of these enzymes is directed to their target RNAs.

5. What implications does the discovery of RNAi have for basic research and disease treatment? How might effective RNAi be more difficult to achieve in humans than in worms?

6. In Fire and Mello's initial RNAi experiment, dsRNA was introduced into the worms by injecting the adult worms and then observing their progeny. Within the next year, it was discovered that RNAi could also be induced in worms either by soaking the worms in a solution containing dsRNA or by feeding the worms *E. coli* bacteria that produce dsRNA and then observing the progeny (*C. elegans* worms normally eat bacteria). (a) How do you think the bacteria were engineered to produce dsRNA? (b) What is the advantage of using this bacteria-feeding strategy? (c) *E. coli* normally produce an endonuclease that is specific for dsRNA (RNaseIII). If you were the researcher trying to optimize RNAi in worms, what would you do next given this information?

Chapter 4 Reading #4 and Discussion Questions

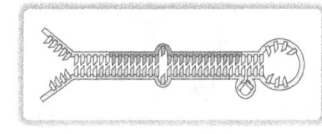

Reading: *Molecular Biology: Concepts for Inquiry*
Chapter 4: Section 4.2C-E

Online Quiz: 4.2C-E

Discussion Questions:

1. What creates the phenotypic differences between humans and less complex animals?

2. What is meant by "noncoding RNA"? How do noncoding RNAs differ from mRNAs?

3. How is your understanding of the nature of genes and genomes different now compared to after completing your first Biology course? Why do you think the information is often oversimplified in early courses?

4. Explain what would be necessary for an RNA molecule to adopt a hairpin shape.

5. Why is RNAi thought to have evolved? What are the advantages to an organism of using miRNAs to regulate the expression of other genes?

6. Explain the two different ways that miRNAs can regulate gene expression. What determines which pathway will be followed?

7. What are the differences and similarities between siRNA, miRNA, shRNA. If you were going to use RNAi to knock down the expression of a gene in a human cell line, which would you use and why? Why would it not be wise to only try to design your dsRNA against one specific segment of the target gene?

8. How are researchers combining the findings of the human genome project with their understanding of RNAi in searching for disease treatments?

9. Explain how the length of *XIST* is important to its function.

Chapter 4 Reading #5 and Discussion Questions

Reading: *Molecular Biology: Concepts for Inquiry*
Chapter 4: Section 4.3

Online Quiz: 4.3

Discussion Questions:

1. What is epigenetic inheritance? How does it differ from genetic inheritance?

2. In some cases, environmental conditions during a person's lifetime have been associated with epigenetic changes in that person's offspring. What kinds of molecular events in what cells would need to occur to explain this observation?

3. How are the different transcriptional patterns in different cell types established? How is the transcriptional pattern in an adult cell passed on to its progeny?

4. Why and in what ways aren't cloned animals completely normal?

5. Why do you think scientists, from a <u>scientific</u> standpoint, almost universally oppose cloning people? Yes, there are significant ethical debates that should be had as well, but what are the problems with cloning humans in terms of the science?

6. Compare and contrast organismal cloning and therapeutic cloning.

7. Explain how the calico cat provides visual evidence of both X-inactivation and issues with organismal cloning.

8. What are the current challenges in stem cell research? What are the advantages and disadvantages of using stem cells to treat diseases?

9. What is the role of lncRNAs in genetic imprinting?

10. In what kinds of situations can epigenetic marks contribute to diseases?

Class Activity 4C: Epigenetic Inheritance

Work with others in your small class group to complete these questions. Thoroughly discuss one question at a time without skipping ahead. Use your logic skills to answer the questions - do not look up answers in your textbook or in another reference.

In a pedigree, circles represent females, squares represent males, and diamonds are either gender. A number in a symbol indicates multiple offspring. Individuals with filled symbols have the disease phenotype. Some parents from outside the family are not shown.

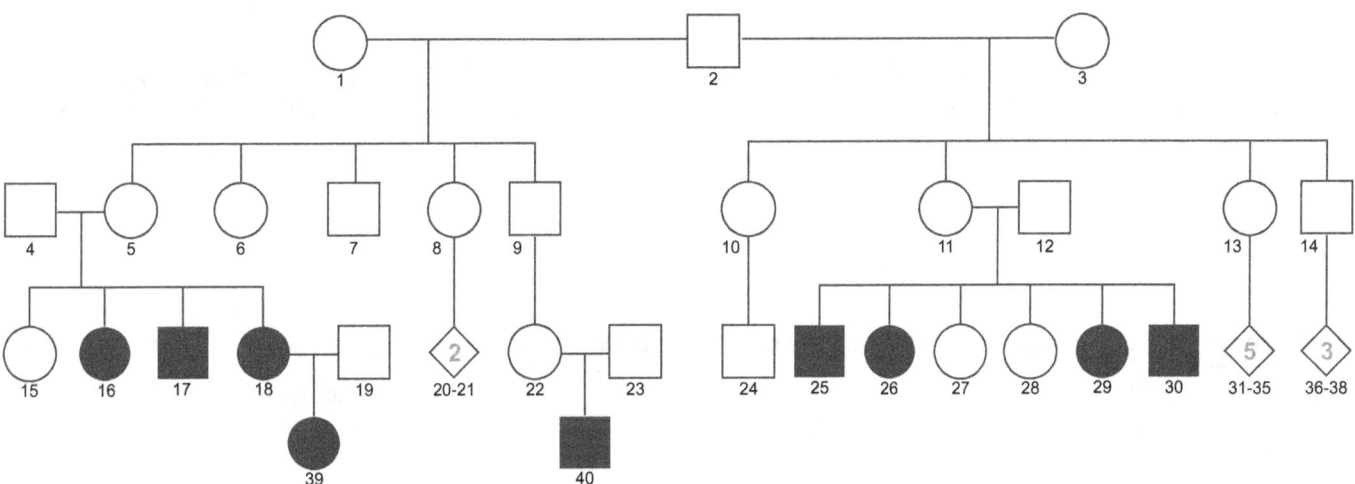

1. Given the family above, what are the possible types of inheritance that could explain the pattern? (Yes, the mutant allele was already present in the first generation).

2. One of the types of inheritance you listed above is technically possible, but very unlikely. Why is it so unlikely? Which of the types of inheritance that you listed above is the most likely and why?

STOP. Check with your instructor before moving ahead. If you only came up with the possible, but unlikely type of inheritance, you should keep thinking and discussing – this takes some logic to figure out.

Beckwith-Wiedemann Syndrome (BWS) is the genetic disorder in the above pedigree. Its symptoms include abnormally large growth in the womb and after the child is born. It is also associated with a predisposition to forming tumors. The genetic locus where the disease gene is located is 11p15.5. This is an example of a locus that exhibits genomic imprinting because the expression of a gene is dependent on the parental origin of the allele. In the case of the above pedigree, the paternal allele is normally silenced and the maternal allele is normally expressed.

3. If you haven't yet done so, indicate the genotypes and imprinting of the alleles for individuals in the pedigree.

4. Sketch in the most likely genotype/phenotype symbols for the parents of person #2 on the pedigree.

5. Describe the possible chromosomes and kinds of mutation (loss/gain of function) that could be responsible for the disease phenotype.

6. If you were the geneticist advising the family, how might you advise different members of the family concerning their risk of having a child or grandchild with Beckwith Wiedemann syndrome? Consider #24, 25, 26, and 36.

7. What type of genetic test would you design to test for the Beckwith-Wiedemann genotype?

8. BWS has an incidence of 1 in about 14000, with about 80% of these cases being caused sporadically by abnormalities in epigenetic marks, rather than by genetic mutations like the mutation in the above pedigree. Multiple types of abnormal epigenetic regulation of multiple genes at the 11p15.5 locus are found in patients, including loss of methylation, gain of methylation, and the presence of 2 chromosomes inherited from one parent (uniparental disomy). There is evidence that the abnormally large size of some cloned animals is associated with changes at the BWS locus. Would you predict that this would be caused by genetic or epigenetic abnormalities? Explain.

9. Compare and contrast imprinting and sex-linked recessive inheritance.

10. Without looking back at the first page, sketch a pedigree for a family that demonstrates the effects of silencing of the <u>maternal</u> allele. Your pedigree should rule out other forms of inheritance, so that if you asked someone else to determine the most likely form of inheritance, the likely answer would be imprinting.

Chapter 4 Reading #6 and Discussion Questions

Reading: *Molecular Biology: Concepts for Inquiry*
Chapter 4: Section 4.4

Online Quiz: 4.4

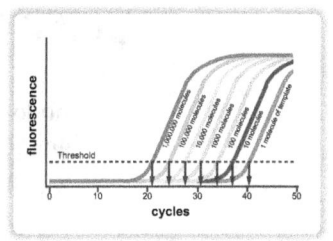

Discussion Questions:

1. Why can't the level of an mRNA be determined from running total cellular RNA on an agarose gel?

2. Compare and contrast northern blotting, qPCR, and microarrays as methods for monitoring mRNA levels.

3. Why is the human proteome much larger than the human genome?

4. Why is it more difficult to measure protein levels than mRNA levels on a genome-wide scale?

5. In what situation(s) would it be beneficial to be able to monitor the entire human transcriptome or proteome? What are the advantages and disadvantages of each approach?

6. If you wanted to determine which genes are transcribed too much or too little in pancreatic cancer cells compared to normal pancreas cells, how would you design experiments to answer this question?

Unit 4 Self-Assessment

1. Rhodopsin is a light-sensitive protein produced in mice (and humans) so that they can perceive light.
    a. For each of the following techniques, how would you predict the amount of rhodopsin in a sample of eye cells would compare to the amount of rhodopsin in the same number of heart cells? Circle the cell type with the HIGHER amount.

        i.   northern blot: eye cells   heart cells   it's the same in both

        ii. Southern blot: eye cells   heart cells   it's the same in both

        iii. western blot: eye cells   heart cells   it's the same in both

        iv. mass spectrometry: eye cells   heart cells   it's the same in both

    b. Circle the following characteristics that are likely to be true of the chromosome at the rhodopsin locus in eye cells:

        i.   methylated promoter   unmethylated promoter

        ii. acetylated histones   histones are not acetylated

        iii. loosely-packed chromatin   tightly-packed chromatin

        iv. euchromatin   heterochromatin

2. A "Golden Gate" assembly of a yeast expression vector is carried out using the following reagents:
    i.   Plasmid containing a yeast promoter between *Bsa*I sites
    ii. Plasmid containing a yeast ORF between *Bsa*I sites
    iii. Plasmid containing a yeast terminator between *Bsa*I sites
    iv. Acceptor vector containing a bacterial *RFP* gene between *Bsa*I sites
    v.   Two enzymes and appropriate buffer.

    a. The "two enzymes" in (v.) are *Bsa*I and_____.

    b. Can the final yeast expression vector be cut by *Bsa*I? Explain.

    c. How is the order and orientation of the DNA fragments determined when the final vector is assembled?

3. State two ways that a prokaryotic gene would need to be altered before it could be expressed in a eukaryotic cell.

4. Describe two different ways that a promoter can play a role in regulating the rate of transcription.

5. Can the transcription of a gene for a microRNA lead to the silencing of its target mRNA if the cell doesn't produce Drosha? Explain.

6. Explain the significance of the length of *XIST* RNA.

7. Why do cloned animals have more health problems than the animals from which they were cloned?

8. Explain the role of charge in the similarities and differences between the packing of euchromatin and heterochromatin. Your answers may include the terms "DNA", "histones," and "acetyl groups."
    a. How charge explains the similarities between their structures:

    b. How charge explains the differences between their structures:

9. Describe the structure of a gene that would cause RNAi in *C. elegans*.

10. How are methylation patterns maintained as cells go through DNA replication and mitosis?

11. Describe one role of the sequence of the 3' untranslated region of mRNAs in the regulation of gene expression in eukaryotes.

12. a. Provide an explanation for the inheritance pattern observed in this pedigree. Explain your reasoning. Squares are males, circles are females. Assume filled shapes indicate the disease phenotype. Assume this genetic disease is <u>extremely</u> rare. Also assume that there are no new mutations occurring in this family.

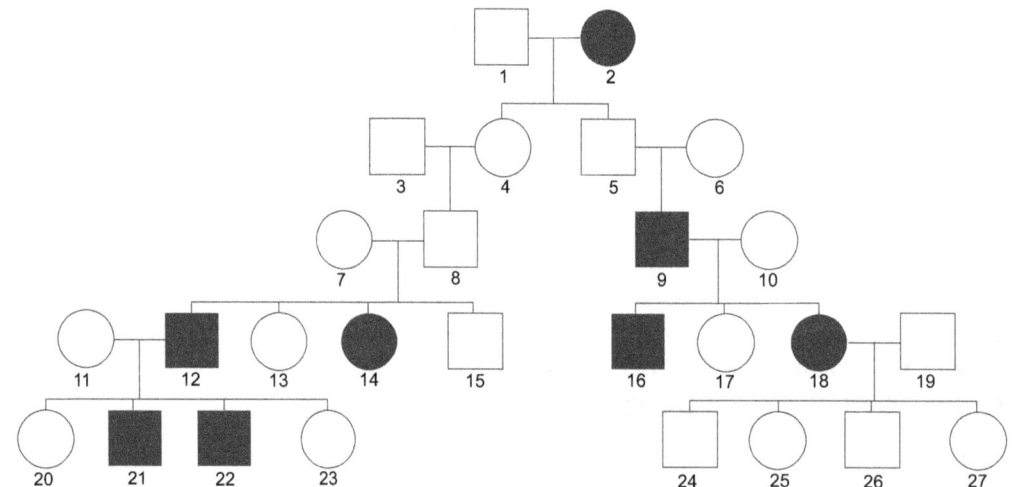

b. What is the chance that a child of person #15 will have the disease?

c. What is the chance that a child of person #26 will have the disease?

Unit 5

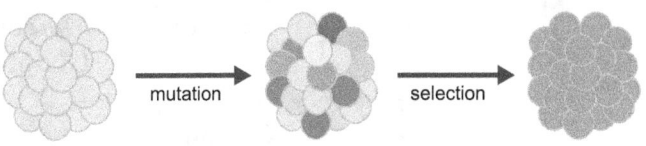

Genome Evolution

Evolution usually requires two processes, mutation and selection. These two processes play key roles in the evolution of chromosomes that occurs both as species evolve and in the "evolution" of cells to form a malignant tumor (cancer). Unit 3 discussed some of the mechanisms that cells have to respond to DNA damage. These processes prevent mutations the majority of the time, but rarely mutations can occur. If mutations occur in somatic cells, they may lead to cancer if they increase cell division or decrease cell death. If mutations occur in germ cells, they may lead to genetic disease or create changes in phenotype that are selected in evolution.

To achieve the deepest level of understanding, you are encouraged to complete the class activities, textbook readings and question sets in the order listed below.

Suggested order:
  Chapter 5 Reading #1 and Discussion Questions
  Class Activity 5A: Cancer Genetics
  Chapter 5 Reading #2 and Discussion Questions
  Chapter 5 Reading #3 and Discussion Questions
  Chapter 5 Reading #4 and Discussion Questions
  Class Activity 5B: Telomeres and Mutation Rate
  [Experiment: Telomeres and Mutation Rate]
  Chapter 5 Reading #5 and Discussion Questions
  Unit 5 Self-Assessment Questions

Prior knowledge (review if necessary):
- Structure of DNA, RNA, and proteins (in Chapters 1, 2. 3)
- Complementary DNA and RNA nucleotides (in Chapters 1, 3)
- How DNA codes for RNA which codes for protein (in Chapter 2)
- DNA replication (in Chapter 3)
- Mutation and DNA repair (in Chapter 3)
- The cell cycle and mitosis (in Chapter 3)
- Enzyme structure and function (in Chapter 2)
- Kinase signaling cascades (in Chapter 2)
- Organization of genes on chromosomes (in Chapter 3)
- Dominant and recessive, loss of function, haploinsufficiency, gain of function, dominant negative (in Chapter 2)
- RNAi (in Chapter 4)
- Epigenome maintenance (in Chapter 4)
- The regulation of gene expression (in Chapter 4)

Chapter 5 Reading #1 and Discussion Questions

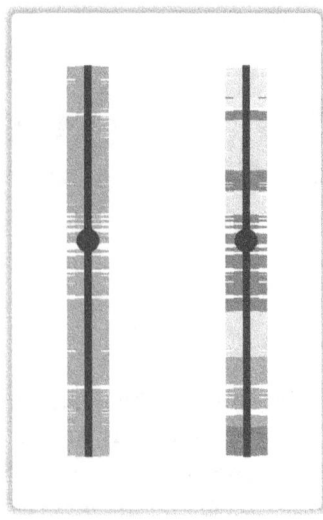

Reading: *Molecular Biology: Concepts for Inquiry*
Chapter 5: Introduction and Section 5.1

Online Quiz: 5.1

Discussion Questions:

1. Explain why both mutation and selection are usually required for evolution. Give an example.

2. What determines whether a part of a genome is subject to natural selection? What are the characteristics of the parts of a genome that are subject to natural selection compared to other parts of a genome?

3. Describe the kinds of changes to the DNA that can occur as genomes evolve.

4. How can the current state of the genomes of two species be used to determine their evolutionary history?

5. What is a block of conserved synteny? What are orthologous proteins? Why does the DNA sequence tend to be more conserved within genes for orthologous proteins than within the rest of a syntenic block? How does the pattern of syntenic blocks reflect the relatedness of species?

6. How does gene duplication (of whole genes or parts of genes) contribute to the evolution of novel genes?

7. Summarize how both mutation and selection contribute to the evolution of genomes.

8. A whole-genome duplication event happens in a species. After a very long evolutionary time, how would you predict the genome might be the same or different than the genome at the time of the whole-genome duplication?

9. Is it reasonable to say that more complex organisms have been evolving longer than less complex organisms? Explain.

10. ⚙ Single-nucleotide polymorphisms (SNPs) are changes in single DNA nucleotides compared to the reference genome sequence. SNPs are the sequences that are analyzed using microarrays by the companies that mass-market general DNA tests of ancestry or disease risk. Given your understanding of genomes, why does it make sense that the vast majority of the 10 million known SNPs are not located in protein-coding sequences? These companies generally focus on analyzing SNPs with high frequencies of variation in the population and often deliberately avoid checking the few SNPs in protein-coding sequences. For example, when the author's DNA was analyzed in one of these commercial tests for a school project, she was heterozygous for 34% of the 130,000 SNP nucleotides that were tested. How do you think the impression that these companies and television programs that use these tests to study celebrities' ancestry create of the amount of DNA variation in the human population compares to the actual amount of variation?

11. ⚙ Explain the effect that meiosis would have on the inheritance of clusters of SNPs relative to how closely the SNPs are located to each other on a chromosome.

12. ⚙ Mitochondria contain their own circular genome. Mitochondria are present in the egg, but are not usually transferred into the zygote from the sperm. Why is mitochondrial DNA or Y chromosome DNA often used to provide evidence of a person's inclusion in a particular family? How would the interpretation of Y chromosome sequence differ from the interpretation of sequence on the other nuclear chromosomes?

13. ⚙ The genes in the human genome were identified in part during the Human Genome Project by comparing the human genome sequence to the mouse genome sequence. What qualities do you think were detected to identify the gene sequences?

Class Activity 5A: Cancer Genetics

Work with others in your small class group to complete these questions. Thoroughly discuss one question at a time without skipping ahead. Use your logic skills to answer the questions - do not look up answers in your textbook or in another reference.

(The videos and the Hit Simulator referenced by URL in this activity were developed by BSCS/NIH for the BSCS/NIH Curriculum Supplement: *Cell Biology and Cancer* to be used with a different set of questions. Some of the teaching goals of that curriculum supplement were the same as some of the goals of this activity. This activity seeks an overall deeper understanding of the genetics of cancer, but portions of this activity were heavily inspired by the BSCS/NIH curriculum supplement, particularly in the use of the videos, the use of data regarding colon cancer incidence with age, and in the overall goals of using the Hit Simulator. The inclusion of pedigree analysis, the analysis of childhood cancers, the comparison of the genetics of childhood and adult cancers, and all of the questions in this activity are novel).

Watch these video clips that depict some of the early evidence that was considered by scientists trying to understand the cause of cancer. (These videos were developed for the BSCS/NIH Curriculum Supplement: *Cell Biology and Cancer*). https://science.education.nih.gov/supplements/webversions/CellBiology/activities/activity2_videos.html

1. Do these videos suggest a common cause for all cancers?

2. What about this information do you think might have initially made it difficult to figure out the cause of cancer?

Timeline: Major discoveries in regard to evolution and heredity
    1859 Darwin publishes *On the Origin of Species*
    1865 Mendel's work is published, but ignored/unknown.
                Genes are units of heredity
                Alleles are different versions of the same gene
                There are 2 alleles for each gene in an individual that may be identical or different
                Dominant and recessive alleles
                Law of segregation
                Law of independent assortment
    1900 Mendel's work is rediscovered.
    1910 Thomas Hunt Morgan proposes that genes are located on chromosomes.
    1911 Sturtevant and Morgan discover that particular genes are arranged in particular orders on particular chromosomes.
    1913 Sturtevant and Morgan discover that one allele can mutate into a different allele. Also, that two different alleles are located at the same position on the two different chromosomes in a pair.
    1944 Avery, McLeod and McCarty show that DNA is the genetic material.
    1953 Watson and Crick determine the structure of DNA.

3. By 1900, it was known that tumors have abnormally high rates of cell division. It was also known that some kinds of cancer run in families. In 1914, Boveri was the first to propose that chromosomal abnormalities might contribute to abnormal rates of cell division in cancer. Given the timeline above, speculate about how previous discoveries might have influenced Boveri's thinking.

4. Retinoblastoma is a cancer of the retina (eye) that affects 1 in 20,000 children. In the following pedigree, affected individuals have retinoblastoma. **Based only on the information provided here**, what would you predict is the most likely mode of inheritance of this disease - dominant or recessive? Why?
(Squares are males, circles are females, filled symbols have the retinoblastoma phenotype).

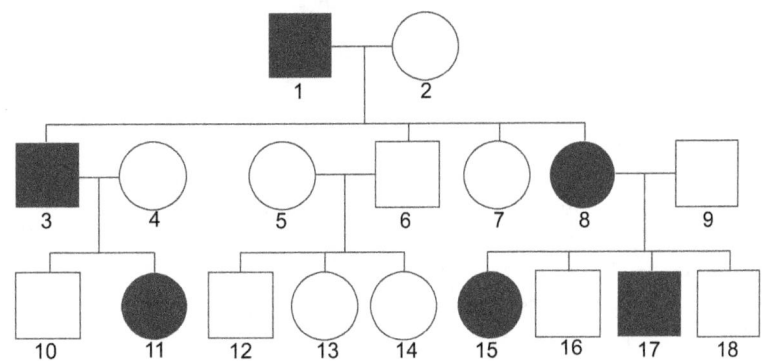

5. When tumor cells from a patient with retinoblastoma are fused with normal cells, the fused cells have a normal phenotype. What does this suggest about whether the allele causing retinoblastoma is dominant or recessive? Explain.

6. Propose an explanation that reconciles your answers to questions 4 and 5 (your answers likely contradicted each other).

7. Given your new explanation, label the genotypes that individuals inherit in this new copy of the pedigree:

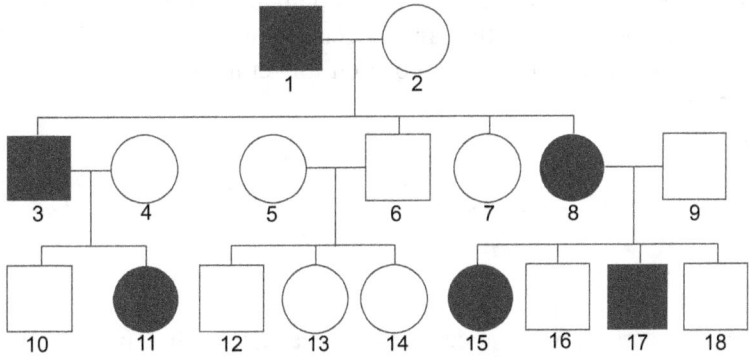

STOP: Check with your instructor before moving on. If you considered haploinsufficient dominance for question 6, good reasoning! But it turns out not to be the explanation – can you come up with another one? Be sure that your new explanation explains all of the data in questions 4 and 5.

Class Activity 5A: Cancer Genetics   Unit 5 | 99

8. How many mutant Rb alleles were inherited by person 8 in the pedigree above?

9. The pedigree below shows the occurrence of retinoblastoma in a different family, but this time indicates whether the patient had retinoblastoma in one eye (unilateral) or in both eyes (bilateral).

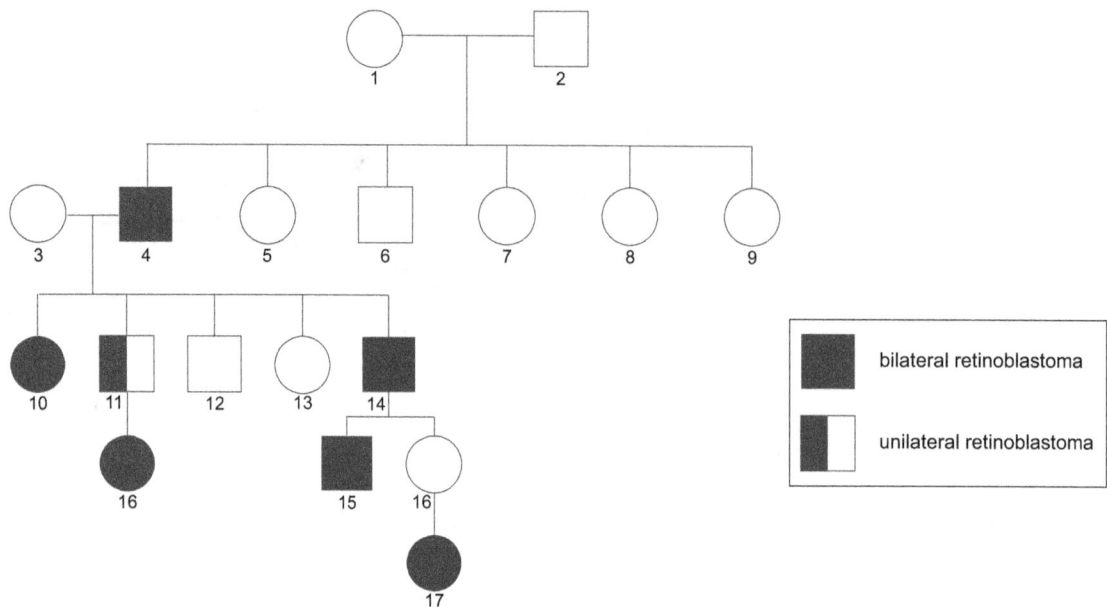

a. Explain how it's possible that some patients have bilateral retinoblastoma, while others in the same family have unilateral retinoblastoma.

b. Explain how it's possible for retinoblastoma to skip a generation (as in the pedigree above).

10, In 40% of cases of retinoblastoma, a predisposition to developing the cancer is inherited (as in the pedigrees above). However, in 60% of cases of retinoblastoma, the cancer occurs sporadically (the first case in a family) and children of these patients do not necessarily inherit a predisposition for retinoblastoma. What is the same about these two situations? What is different?

11. In Chronic Myelogenous Leukemia (CML; cancer of blood precursor cells), patients have a translocation between chromosomes 9 and 22 (meaning that these two chromosomes swap part of their chromosome arms). On one of the fused chromosomes, part of a gene called *BCR* becomes fused to a gene called *ABL*, resulting in a new protein, "BCR-Abl." Only one copy of this fused gene is present in patients. Putting one copy of the fused human ORF (behind a bone marrow-specific promoter) into mice causes leukemia in the mice and these leukemia cells only have one copy of the *BCR-ABL* gene.

a. How many alleles must be mutant to cause CML?

b. Is the mutant *BCR-ABL* allele dominant or recessive to the wildtype *ABL* allele?

c. How many *RB* alleles must be mutant to cause retinoblastoma?

d. Is the mutant *RB* allele dominant or recessive to the wildtype *RB* allele?

12. By 1914, Boveri had observed that sometimes cancer cells had extra copies of a chromosome and sometimes they had too few copies of a chromosome. Boveri suggested that two different types of chromosome abnormalities (one type with chromosomes present in extra copies, and one type with chromosomes lost) might contribute to causing the high rate of tumor cell division. Note that today, we understand that it's really individual genes, not whole chromosomes that are significant.

    If you haven't read about the regulation of cell division and genes that contribute to cancer in the textbook yet, watch animations 1 and 3: at https://science.education.nih.gov/supplements/webversions/CellBiology/activities/activity2_animations.html or at https://youtu.be/1VGPT9iuzkY (cell division and tumors), and https://youtu.be/2jb89T9Hkrc (tumor suppressor genes and oncogenes).

    For each description below, state whether it describes tumor suppressor genes or oncogenes.

    a. more likely to have gain-of-function mutations in tumors

    b. more likely to have loss-of-function mutations in tumors

    c. wildtype alleles are often present in multiple copies in tumors

    d. wildtype alleles are often lost in tumors

13. Knudson's two-hit hypothesis is that both alleles of a gene must acquire mutations in order for that gene to contribute to cancer.
    a. Do you think that Knudson's two-hit hypothesis applies to tumor-suppressor genes, oncogenes, or both?

    b. Do you think it is more likely that a tumor suppressor gene or oncogene is mutated in:

        i. Retinoblastoma

        ii. CML

    c. With which of the above kinds of patients do you think Knudson worked? Why?

14. Most cancers (90%) do not arise from a familial predisposition and most of these tend to occur at higher rates later in life. For example, the rate of diagnosis of new cases of colon cancer increases with age as shown below:

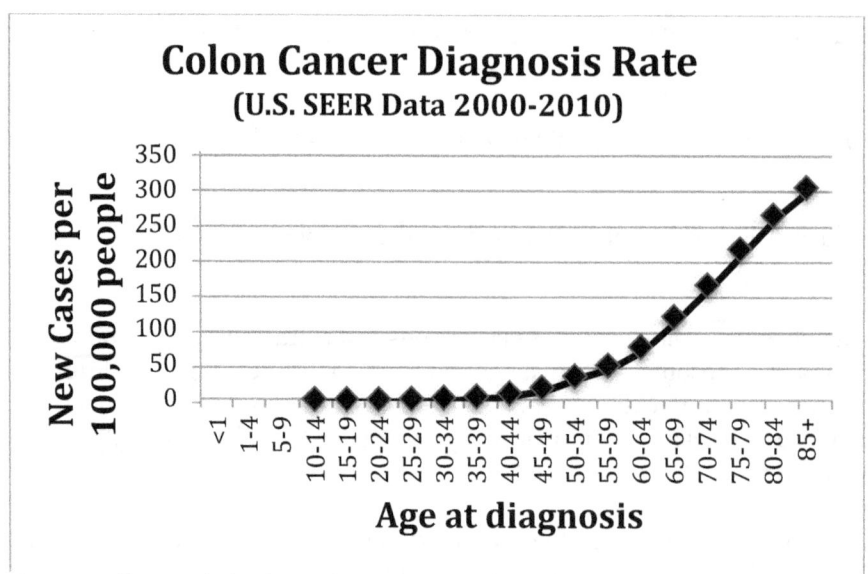

Source: http://seer.cancer.gov/faststats/selections.php?series=cancer.
An earlier set of these data were used in the BSCS/NIH curriculum supplement.

By observing the data in the graph above,
   a. speculate about possible reasons why colon cancer does not usually occur in young people.

   b. Do you need any more information in order to predict how many genes must be mutated to cause colon cancer? Explain.

15. Open the spreadsheet "Activity 5A_CancerMutationSimulator" at https://hackettmolecularbiology.blogspot.com/ and make a copy of the spreadsheet that you can edit. (The calculations performed by this spreadsheet are the same as those performed by the "Hit Simulator" at https://science.education.nih.gov/supplements/webversions/CellBiology/activities/activity3.html, but this new version allows a more straightforward comparison to the data in the graph above). This simulator will allow you to figure out approximately how many genes must be mutated to cause colon cancer. In the spreadsheet, the "number of genes that must be mutated to cause cancer (1-7)" can be altered (yellow box) and it assumes that hits have occurred in both alleles of a tumor suppressor gene or in one allele of an oncogene. The "percent of people in each age interval who get mutations in a particular gene (0-1)" can also be altered (blue box). For example, if "number of genes" is set to "2," then mutations in 2 genes are necessary to create a cancer. In this scenario, if "percent of people in each age interval" is set to 0.5, then by age 5, 50% of people will get mutations in gene #1 and 50% of people will get mutations in gene #2, so 25% of the people will get mutations in both genes and have cancer. Figure out how this simulator works before moving on by changing the values in the yellow and blue boxes.

16. Change the values in the simulator until the "simulated" plot in the simulator approximates the "actual" plot, which is the same as in the graph above. Based on your observations, approximately how many different genes must be mutated in order for someone to develop colon cancer? Explain how the simulation data indicate this. Is there any uncertainty in your estimate? Explain.

17. The following graph shows the incidence of retinoblastoma by age:

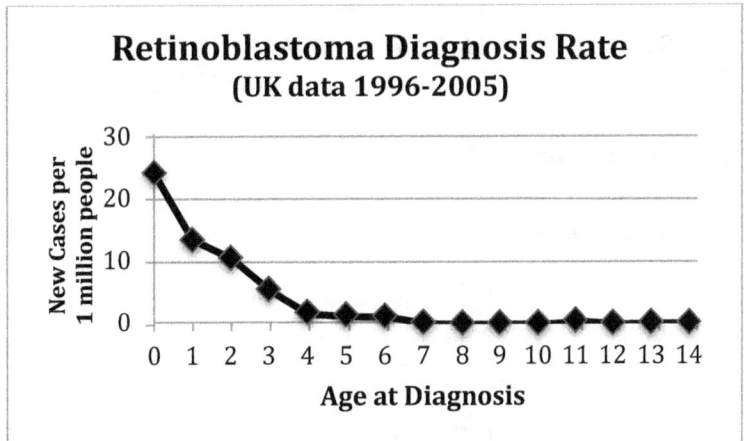

Source: http://www.cancerresearchuk.org/cancer-info/cancerstats/childhoodcancer/incidence/childhood-cancer-incidence-statistics

Additional information about retinoblastoma:
- Retinoblastoma develops in fetal retinoblast cells.
- Fetal retinoblasts normally develop into retinal epithelial cells (the eye cells that detect light).

   a. Compare retinoblastoma to colon cancer. Do the different shapes of the graphs for diagnosis rate for these two cancers imply that they arise through different basic mechanisms? How does this fit with the information about the cells of origin for retinoblastoma described above? How can you reconcile this information into one unifying explanation for how cancer arises?

   b. In what ways are embryonic/fetal cells and cancer cells similar?

   c. Speculate about what role the cells in which cancer originates might play in terms of influencing the age of developing cancer and the number of gene mutations required to cause cancer.

   d. In general, the rate of curing childhood cancers with treatment is higher than the rate of curing adult cancers. How might this make sense in terms of the relative genetic complexity in childhood cancers and in adult cancers?

The following article is recommended to explore these topics more, and the article includes a detailed timeline of significant events in understanding how cancer arises:
Knudson, A. G. Two Genetic Hits (More or Less) to Cancer. *Nature Reviews Cancer* 1, 157-162 (2001).

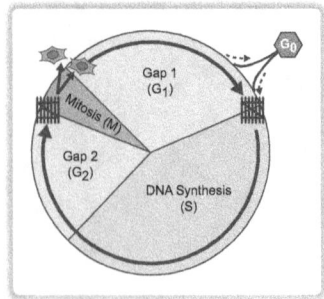

Chapter 5 Reading #2 and Discussion Questions

Reading: *Molecular Biology: Concepts for Inquiry*
Chapter 5: Section 5.2A-E

Online Quiz: 5.2A-E

Discussion Questions:

1. Why is cancer said to be a collection of genetic diseases? Can cancer be inherited?

2. Explain the roles of mutation and selection in tumor formation. How is this similar to and different from what occurs in genome evolution during the evolution of species?

3. How do cell cycle checkpoints and apoptosis help to prevent tumor formation?

4. Explain the roles of tumor suppressor genes, oncogenes, and DNA repair genes in tumor formation.

5. Examine Figure 5.8. Why does it make sense that usually one gene per pathway is mutated in a tumor?

6. Why does it make sense that *p53* is one of the genes that is most frequently inactivated in tumors? (How does it differ from other cancer genes)?

7. Why does it make sense that kinases are often oncogenes?

8. Explain Knudson's two-hit hypothesis. How many alleles of a tumor suppressor are usually mutated? How many alleles of an oncogene are usually mutated? If a patient has retinoblastoma tumors in both eyes, give the most likely genetic explanation for their disease.

9. Why do retinoblastoma and osteosarcoma usually occur in childhood, but not often later in life?

Chapter 5 Reading #3 and Discussion Questions

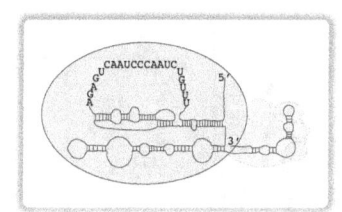

Reading: *Molecular Biology: Concepts for Inquiry*
Chapter 5: Section 5.2F-H

Online Quiz: 5.2F-H

Discussion Questions:

1. What are the roles of angiogenesis and metastasis in cancer?

2. Explain the "end replication problem." How does telomerase elongate telomeres?

3. How does the biology of telomeres explain both the Hayflick limit (cellular senescence) and cellular immortality?

4. In which cells is telomerase expressed? Why do we think telomerase isn't expressed in all cells?

5. How do telomeres distinguish natural DNA ends from DNA breaks?

6. What is meant by "survival of the fittest?" How does cancer affect the fitness of individual cells? The organism? How has this contributed to the evolution of tumor suppressor genes/processes?

7. Considering the characteristics that tend to be common to all cancers in Table 5.4, why does it make sense that all cancers would need to have these characteristics?

8. Mice that always express telomerase have higher rates of developing cancer. Explain why this makes sense.

9. Before telomerase's role in the Hayflick limit was understood, the kinds of human cell lines that were used in labs for tissue culture experiments were almost all derived from human tumors (sometimes without the informed consent of the patient, as in the case of Henrietta Lacks' cells that were used to create the first immortal cell line). In what ways might it be problematic that a lot of research is based on studies using cancer cells?

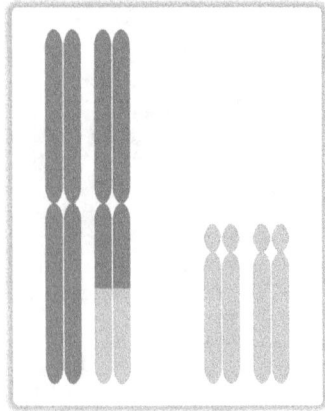

Chapter 5 Reading #4 and Discussion Questions

Reading: *Molecular Biology: Concepts for Inquiry*
Chapter 5: Section 5.2I-L

Online Quiz: 5.2I-L

Discussion Questions:

1. Explain how the types of mutations that occur in cancer can lead to either the loss of function of tumor suppressor genes or to the gain of function of oncogenes.

2. Why does it make sense that the few cancers without genomic instability display epigenomic instability?

3. Compare and contrast telomere shortening and chromosome missegregation as causes of genomic instability in cancer. Are these possibilities mutually exclusive?

4. Why do you think it's difficult to determine the exact cause of genome instability in cancer?

5. Once the mutation rate in tumor cells becomes elevated, what kinds of strategies might a cell use to reduce the mutation rate?

6. Explain how tetraploidy can contribute to tumor development.

7. Explain the role of selection in strategies to treat cancer.

8. How can studies of protein structure lead to improved cancer treatments?

Class Activity 5B: Telomeres and Mutation Rate

Work with others in your small class group to complete these questions. Thoroughly discuss one question at a time without skipping ahead. Use your logic skills to answer the questions - do not look up answers in your textbook or in another reference.

Students set out to study the relationship between the loss of telomere end-protection and mutation rate. They did so by conducting experiments in yeast that are simplified versions of the experiments in the following references:
Hackett, J. A., Feldser, D. M. & Greider, C. W. Telomere dysfunction increases mutation rate and genomic instability. *Cell* 106, 275-286 (2001).
Hackett, J. A. & Greider, C. W. End resection initiates genomic instability in the absence of telomerase. *Molecular and Cellular Biology* 23, 8450–8461 (2003).

Complete the following activity as if you and your classmates are preparing to do the experiments and then obtain the data that is described.

PART 1: Pre-lab

Purpose: To measure the frequency of mutation and calculate the mutation rate in wildtype yeast and yeast with rapidly-shortening, unprotected telomeres (heat-shocked *cdc13-1*).

Yeast Nomenclature:

Wildtype gene name is capital, italicized:	*CDC13; CAN1*
Mutant gene name is lowercase, italicized:	
Δ: deleted gene:	*can1Δ*
gene name-number: indicates a specific (usually point) mutation in the gene:	*cdc13-1*
Protein name is normal font, first letter is capitalized, and ends in "p":	Cdc13p; Can1p
Sensitive or resistant phenotype is indicated with an "S" or "R" superscript. The wildtype phenotype is uppercase and the mutant phenotype is lowercase:	CAN^S; can^R
Haploid yeast have a single name for each gene:	*LYS5* or *lys5*
For diploid yeast, the names of both alleles are written:	*lys5/LYS5* or *lys5/lys5* or *LYS5/LYS5*

Background: Mutations occur at a low rate in all living things. One type of mutation, a point mutation, is a substitution, insertion, or deletion of a single nucleotide in the DNA that changes the amino acid sequence of a protein. Other types of mutations, gross chromosomal rearrangements, include the deletion of large sections of DNA or even loss of a whole chromosome. The mutation rate is elevated in cancer cells and a question is what might cause this increased rate of mutation, particularly the increase in the rate of gross chromosomal rearrangements that is commonly observed in tumors. One hypothesis is that short telomeres can increase the rate of formation of gross chromosomal rearrangements. We will study this possibility in this experiment using yeast as a model system

Yeast are simple single-celled eukaryotes, so they have linear chromosomes and telomeres that are maintained by telomerase. Yeast divide rapidly (every 1.5 hours) and so are useful for mutation rate experiments in which we want our cells to go through many cell divisions in a short period of time. Like bacteria, yeast can be grown in liquid cultures or in colonies on plates. Yeast can be haploid or diploid. Our diploid yeast have two copies of each of their 16 chromosomes.

The yeast we are using have been genetically engineered so that both point mutations and large deletions can be detected. Normally, yeast die when they are exposed to a toxic drug called **canavanine**. Any mutation in the *CAN1* gene will allow the yeast to grow on mutant-selection plates containing the drug canavanine. Wildtype *CAN1* yeast are canavanine-sensitive (CAN^S) and mutant *can1* yeast are canavanine-resistant (can^R). Therefore, canavanine selects for mutant cells. This works because Can1p is an arginine permease – it is a cell membrane transport protein that normally transports the amino acid arginine into the yeast cells. Since canavanine is a close molecular analog of arginine, wildtype Can1p will also transport canavanine into the cells, which then kills them. However, any mutations that inactivate Can1p will prevent the transport of canavanine into cells and allow their growth on plates containing canavanine. It's OK that this also prevents the transport of arginine because yeast can synthesize their own arginine so they don't really need to transport it.

The yeast we will be using are red (wildtype yeast are white) because they are homozygous for mutations in the *ADE2* gene and so have an *ade2/ade2* genotype. *ade2/ade2* yeast are red because Ade2p is an enzyme required for the synthesis of the nitrogenous base adenine and the loss of Ade2p blocks the process of synthesizing adenine at a step where a red pigment accumulates in the cells (Figure 1a,b). Since the synthesis of adenine is blocked, these yeast must be grown on media containing adenine. However, these red yeast can become white if they acquire mutations in another gene in the adenine biosynthesis pathway that functions earlier in the pathway, *ADE5*, since the loss of Ade5p prevents the accumulation of the red pigment (Figure 1c).

Figure 1:

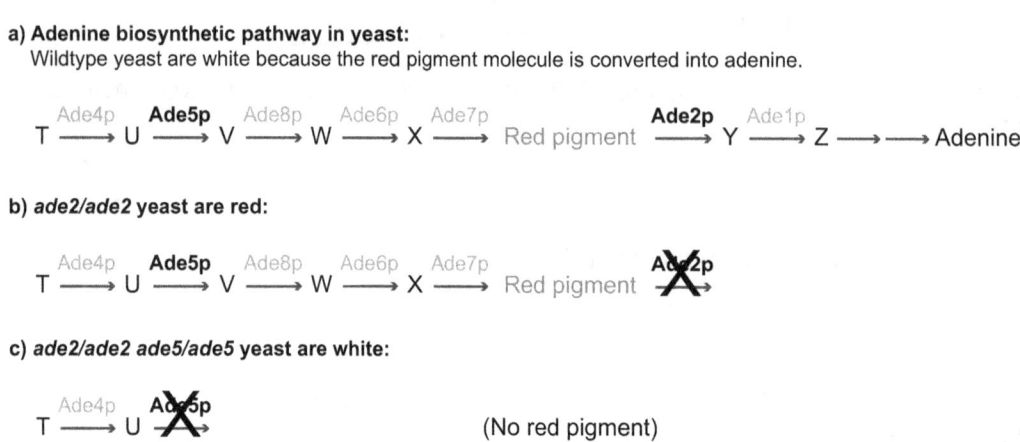

Our yeast have been genetically engineered so the *CAN1* gene is located near the end of one copy of their chromosome VII, but it is missing from the other copy of chromosome VII (Figure 2a). Therefore, only one mutation in *CAN1* is required for mutant yeast to grow on canavanine mutant-selection plates. The yeast have been genetically engineered so *ADE5* is located immediately next to *CAN1* on one copy of chromosome VII (Figure 2a). If there is a point mutation in *CAN1*, the can^R colonies on the canavanine mutant-selection plate will stay **red** (Figure 2b). However, if there is a large deletion or loss of the whole chromosome, the yeast will lose both the red-color *ADE5* gene and *CAN1*, so the *ade5* can^R colonies on the canavanine mutant-selection plate will be **white** (Figure 2c).

Figure 2:

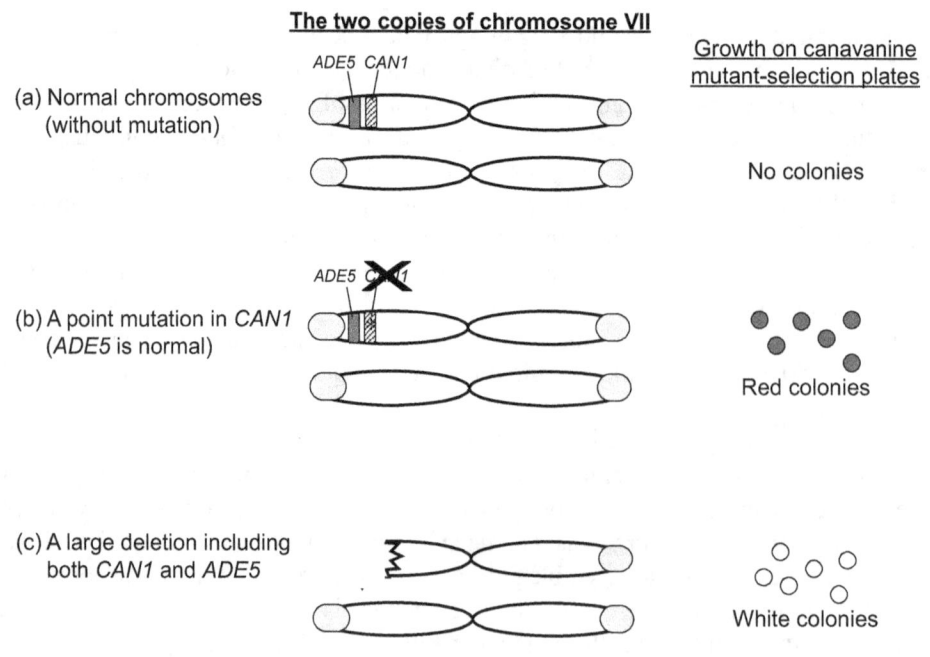

1. If you get a colony that grows on the drug canavanine, what do you know?

2. What will be observed when each of the following kinds of yeast cells are plated on canavanine mutant-selection plates?
   a. Normal yeast

   b. Yeast with a point mutation in *CAN1*

   c. Yeast with a point mutation in *ADE5*

   d. Yeast with a large deletion that includes both *CAN1* and *ADE5*

   e. Yeast that have lost the copy of chromosome VII containing *CAN1* and *ADE5*

All of our yeast have been genetically engineered to detect mutations as described above. We have two subtypes of these yeast: wildtype and *cdc13-1*. The wildtype yeast are normal. The *cdc13-1* yeast have a mutation in the gene that encodes a telomere-binding protein, Cdc13p. This Cdc13p normally protects the telomere from degradation by exonucleases by acting like the plastic piece (aglet) on the end of your shoelace that keeps the ends from fraying (Figure 3b). Without functional Cdc13 protein, the ends of the chromosome can be degraded by exonucleases (Figure 3c). This leads to large deletions and sometimes chromosome loss. This can also lead to an arrest in the cell cycle or cell death due to the shortening of telomeres. (*cdc13-1* yeast were discovered in a screen for c̲ell d̲ivision c̲ycle mutants that stop dividing. Reference: Hartwell, L.H., *et al.* Genetic control of the cell-division cycle in yeast. I. Detection of mutants. *PNAS* 66, 352-9 (1970)).

The mutation in *cdc13-1* yeast is a temperature-sensitive point mutation, a mutation that makes the protein denature at higher temperatures. The Cdc13 protein in these *cdc13-1* yeast folds properly and functions normally at lower temperatures, including room temperature (22°C) (Figure 3b). However, when heat-shocked by growing at higher temperatures (like 37°C), the Cdc13 protein in *cdc13-1* yeast unfolds and no longer functions (Figure 3c).

Figure 3:

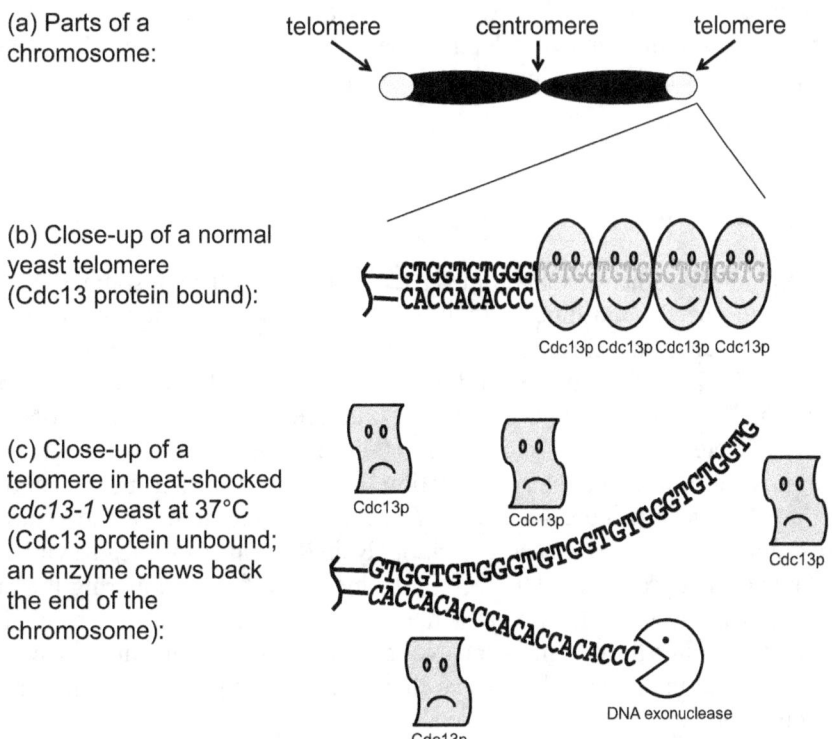

(a) Parts of a chromosome:

(b) Close-up of a normal yeast telomere (Cdc13 protein bound):

(c) Close-up of a telomere in heat-shocked *cdc13-1* yeast at 37°C (Cdc13 protein unbound; an enzyme chews back the end of the chromosome):

3. What is the normal function of Cdc13 protein?

4. What will be the effect on the chromosomes of heat-shocking *cdc13-1* yeast at 37°C?

5. Why will the heat-shock cause the effects you listed above?

6. What should be the effect of heat-shocking wildtype yeast at 37°C?

7. Write a hypothesis predicting how heat-shocking the cells should affect the rates of point mutations and gross chromosomal rearrangements in wildtype and *cdc13-1* yeast.

8. After we heat-shock the *cdc13-1* yeast at 37°C, we will grow them for several hours at room temperature (22°C) before we plate them on canavanine mutant-selection plates. Why do you think this delay is necessary before exposing the yeast to the toxic drug canavanine?

It's beneficial to use diploid yeast rather than haploid yeast to study large chromosomal deletions since diploid yeast can lose up to an entire copy of one chromosome, but still survive if they retain the second copy. Haploid yeast die if large deletions include essential genes on a chromosome. The yeast we are using have been engineered further to allow the easy determination of how far large deletions extend into a chromosome. Figure 4a shows the normal structures of the two copies of chromosome VII and Figure 4b shows the structure if a point mutation occurs in *CAN1*. If a gross chromosomal rearrangement deletes both *ADE5* and *CAN1*, a "large" deletion that just deletes the sequence closest to the telomere can be distinguished from a "very large" deletion that extends further into the chromosome based on the presence or absence of the *LYS5* gene (Figure 4c,d). The *LYS5* gene is required for the synthesis of the amino acid lysine. *LYS5/lys5* heterozygotes can grow on media lacking lysine because they can synthesize their own lysine. However, *lys5Δ/lys5* mutants cannot grow on media lacking lysine. Therefore, comparing the rate at which *ade5* can^R mutants can grow on plates lacking lysine to the rate at which *ade5* can^R mutants require lysine in the plates will tell us how frequently deletions extend far into the chromosome. Notice that it's important that the second copy of chromosome VII contains all mutant versions of the indicated genes so that loss of these genes from the first copy of chromosome VII will produce a mutant phenotype.

Figure 4:

a) Normal structures of the two copies of chromosome VII in our yeast:
ADE5/ade5 CAN1/can1 LYS5/lys5 LEU1/leu1

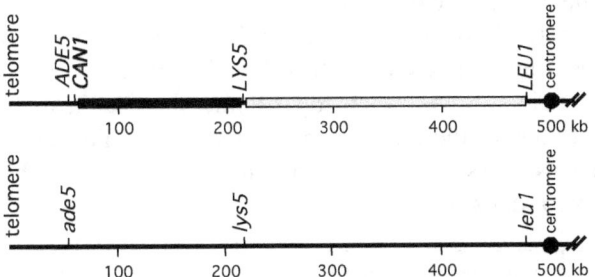

c) Structures of the two copies of chromosome VII in yeast with a large deletion that includes ADE5 and CAN1, but not LYS5:
ade5Δ/ade5 can1Δ/can1 LYS5/lys5 LEU1/leu1

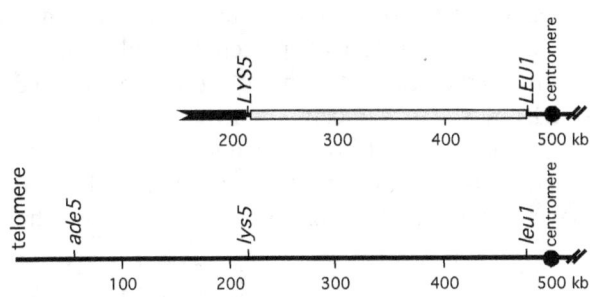

b) Structures of the two copies of chromosome VII in yeast with a point mutation in CAN1:
ADE5/ade5 can1/can1 LYS5/lys5 LEU1/leu1

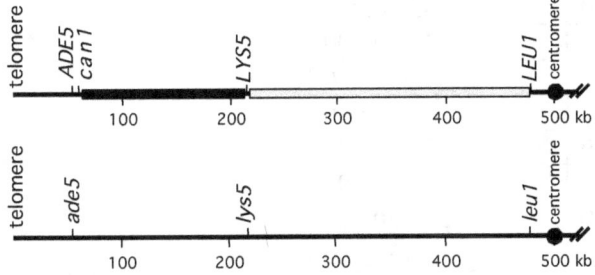

d) Structures of the two copies of chromosome VII in yeast with a large deletion that includes ADE5, CAN1, and LYS5:
ade5Δ/ade5 can1Δ/can1 lys5Δ/lys5 LEU1/leu1

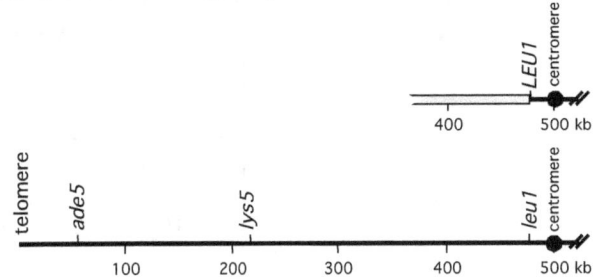

9. Describe the plates you would need to use to select for *can1/can1 LYS5/lys5* colonies.

PART 2: Data Analysis

Analyze the data below as if you performed the experiments as described here. [The data were obtained by J. Hackett's students conducting a class experiment and the data were not necessarily collected with the accuracy and precision that would be expected for publication-quality work. Some of the data that appears to be outlying might have resulted from student error, but overall, the calculated mutation rates are in the expected ballpark].

Each group of students picked one independent wildtype yeast colony and one independent *cdc13-1* yeast colony, inoculated the yeast into 5 mL of liquid growth media at a concentration of 4×10^4 cells/mL and grew the yeast at 22°C for 21 hours. This concentration was chosen so that the yeast would still be in their logarithmic phase of growth at 21 hrs. At the end of 21 hours, each group used a hemocytometer to determine the concentration of yeast. You can notice that all of the cultures were still in the log phase of growth because these cell concentrations were much less than the concentration of yeast during stationary phase, which is over 1×10^8 cells/ml.

Group	[WT yeast] (cells/ml)	[*cdc13-1*] (cells/ml)
1	7,500,000	8,000,000
2	6,500,000	10,000,000
3	10,750,000	5,750,000
4	6,000,000	8,250,000
5	8,750,000	13,500,000
6	9,500,000	9,750,000
7	15,000,000	7,500,000
8	13,000,000	7,250,000
9	10,250,000	11,750,000
10	27,500,000	24,250,000
11	29,000,000	13,750,000
12	21,500,000	22,500,000

1000 yeast of each type were plated on control SC-arginine plates. 200,000 yeast of each type were plated on SC-arginine, -leucine, +canavanine plates and on SC-arginine, -leucine, -lysine, +canavanine plates. All of the plates were grown at 22°C for 3 days.

Two new 5 mL liquid cultures of each cell type were inoculated with 5×10^6 cells/mL (or as high a concentration as possible given the available yeast) and were either heat-shocked at 37°C for 3.5 hours or were grown at 22°C for 3.5 hours. Then all liquid cultures were grown at 22°C for an additional 3.5 hours. At the end of the 7 hours, each group used a hemocytometer to determine the concentration of yeast:

Group	Initial [WT yeast] (cells/ml) in both cultures	Final [WT yeast] (cells/ml) 22°C	Final [WT yeast] (cells/ml) 37°C	Initial [*cdc13-1*] (cells/ml) in both cultures	Final [*cdc13-1*] (cells/ml) 22°C	Final [*cdc13-1*] (cells/ml) 37°C
1	3,600,000	28,000,000	28,000,000	4,000,000	24,000,000	8,250,000
2	1,495,000	35,000,000	53,000,000	5,000,000	13,500,000	12,500,000
3	5,000,000	104,000,000	42,500,000	2,357,500	6,500,000	6,500,000
4	2,760,000	33,000,000	45,000,000	2,277,000	14,250,000	2,500,000
5	3,675,000	26,000,000	42,000,000	5,000,000	40,000,000	8,250,000
6	4,655,000	81,000,000	56,000,000	4,777,500	21,750,000	11,750,000
7	5,000,000	52,000,000	76,000,000	3,900,000	31,000,000	6,250,000
8	5,000,000	35,000,000	58,000,000	2,610,000	30,000,000	6,500,000
9	5,000,000	44,000,000	65,000,000	5,000,000	20,000,000	5,500,000
10	5,000,000	58,500,000	36,000,000	5,000,000	30,000,000	5,000,000
11	5,000,000	33,500,000	62,000,000	5,000,000	11,000,000	5,250,000
12	5,000,000	58,000,000	64,000,000	5,000,000	45,000,000	11,000,000

1000 yeast of each type were plated on control SC-arg plates. 200,000 yeast of each type were plated on SC-arg, -leu, +canavanine plates and on SC-arg, -leu, -lys, +canavanine plates. All of the plates were grown at 22°C for 3 days.

After 3 days, students counted the number of colonies on each plate (the numbers of red colonies on canavanine plates are not listed here because their numbers were too low to calculate point mutation rates):

Group	WT, 0 hrs colonies on control plates	WT, 0 hrs white col. on -leu+can	WT, 0 hrs white col. on -leu-lys+can	cdc13-1, 0 hrs colonies on control plates	cdc13-1, 0 hrs white col. on -leu+can	cdc13-1, 0 hrs white col. on -leu-lys+can
1	3132	42	5	272	50	19
2	609	59	21	532	132	3
3	364	19	2	270	524	9
4	176	29	4	230	82	9
5	114	33	11	162	1254	4
6	436	89	55	186	704	121
7	245	57	27	255	524	104
8	304	54	23	38	35	5
9	225	89	23	280	112	79
10	16	5	1	78	26	6
11	388	54	18	328	19	22
12	1,816	472	278	67	408	420

Group	WT, 7 hrs, 22°C colonies on control plates	WT, 7 hrs, 22°C white col. on -leu+can	WT, 7 hrs, 22°C white col. on -leu-lys+can	cdc13-1, 7 hrs, 22°C colonies on control plates	cdc13-1, 7 hrs, 22°C white col. on -leu+can	cdc13-1, 7 hrs, 22°C white col. on -leu-lys+can
1	472	40	12	507	75	41
2	712	23	24	444	140	85
3	679	43	1	92	308	5
4	968	101	70	328	78	5
5	348	88	16	356	1206	31
6	126	183	35	151	203	122
7	354	36	13	307	520	16
8	136	20	1	359	139	52
9	480	1	25	708	180	179
10	464	33	3	772	286	13
11	400	128	10	516	249	47
12	620	98	13	904	220	22

Group	WT, 7 hrs, 37°C colonies on control plates	WT, 7 hrs, 37°C white col. on -leu+can	WT, 7 hrs, 37°C white col. on -leu-lys+can	cdc13-1, 7 hrs, 37°C colonies on control plates	cdc13-1, 7 hrs, 37°C white col. on -leu+can	cdc13-1, 7 hrs, 37°C white col. on -leu-lys+can
1	656	52	25	137	1152	768
2	620	88	16	41	168	91
3	295	118	6	728	58	4
4	364	239	120	109	640	728
5	514	64	46	144	642	488
6	120	82	18	258	676	119
7	278	20	14	124	520	16
8	239	24	1	20	217	10
9	258	75	17	64	71	194
10	284	98	13	300	1666	629
11	384	86	11	256	960	364
12	416	75	19	320	1428	1264

Class Activity 5B: Telomeres and Mutation Rate

You will use all of these data together to calculate the mutation rate in mutations/cell division. You might have noticed when examining the data above that the number of colonies on different canavanine plates varies widely. This is expected. As shown in the diagram below, a single mutation event during the growth of a culture will result in different frequencies of mutant cells, depending on when that mutation occurred. If the mutation occurs earlier in the growth of the culture, a larger proportion of the cells will be mutant at the end of the growth of the culture. It's necessary to include many separate cultures in an analysis of mutation rate because the time during the growth of a culture at which a mutation occurs will affect the frequency of mutant cells when they are plated:

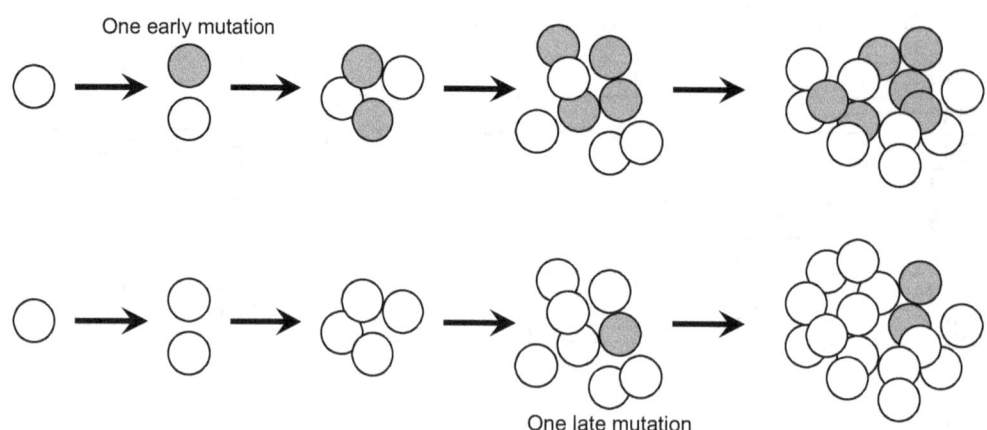

Fluctuation analysis is used to compute the average number of initial mutation events in a set of independent cultures grown in parallel. An initial study of mutation rate by Luria and Delbruck (*Genetics*, 1943) was extended by Lea and Coulson (*J. of Genetics*, 49, 264-285 (1949)) who developed several mathematical methods for calculating mutation rate. These methods are collectively called "fluctuation analysis" because they derive the mutation rate from the data for a large number of parallel cultures in which the frequency of mutant cells fluctuates between cultures.

We will use Lea and Coulson's Maximal Likelihood Method to calculate mutation rate. You can analyze the data yourself using a spreadsheet provided at https://hackettmolecularbiology.blogspot.com/ in which all of the data have been entered. The explanation of how to determine mutation rate below will include directions for how to use the spreadsheet. The details of mutation rate calculation are included here partly to give you an example of the data analysis employed in laboratory research and partly to give you some context for why you are taking the described steps when you calculate a mutation rate from multiple independent cultures.

Maximal Likelihood Method of Lea and Coulson:

Lea and Coulson used calculus to derive equations to calculate the most likely average number of initial mutation events (m) that could have produced the observed number of new mutant cells in each culture at the end of the growth period (r). Lea and Coulson also derived algebraic equations from the calculus that allow m to be determined without actually using calculus and these equations are shown below. The reasoning behind the derivation of these equations will not be described here, but it is explained in Lea and Coulson, 1949.

An initial estimate of the value of m is based on the median value for r, r_0, using the following formula:

$$\frac{r_0}{m} - \ln m = 1.24$$

This formula is used to generate the values in the following table, from which an initial value for m can be determined from r_0:

Table for estimating m from the median, r_0.

r_0	m	r_0	m	r_0	m	r_0	m
1.4	1.06	190	38.9	12,089	1422	629,830	52,052
1.6	1.17	215	42.9	13,518	1572	701,823	57,526
1.9	1.30	242	47	15,113	1737	781,992	63,577
2.3	1.43	273	52	16,895	1920	871,261	70,263
2.7	1.58	307	58	18,884	2122	970,657	77,653
3.2	1.75	346	64	21,104	2345	1,081,324	85,819
3.7	1.93	389	71	23,583	2592	1,204,532	94,845
4.3	2.14	438	78	26,349	2864	1,341,696	104,820
5.0	2.36	493	86	29,437	3165	1,494,388	115,844
5.7	2.61	554	96	32,883	3498	1,664,357	128,027
6.6	2.89	623	106	36,728	3866	1,853,548	141,492
7.7	3.19	700	117	41,018	4273	2,064,125	156,373
8.8	3.53	787	129	45,804	4722	2,298,493	172,819
10.1	3.90	884	143	51,143	5219	2,559,327	190,995
11.6	4.31	993	158	57,099	5768	2,849,601	211,082
13.3	4.76	1115	174	63,741	6374	3,172,625	233,281
15.3	5.26	1251	192	71,149	7044	3,532,074	257,816
17.4	5.8	1404	213	79,411	7785	3,932,039	284,930
19.9	6.4	1575	235	88,623	8604	4,377,064	314,897
22.7	7.1	1767	260	98,894	9509	4,872,206	348,015
25.9	7.8	1981	287	110,346	10,509	5,423,082	384,616
29.5	8.7	2221	317	123,113	11,614	6,035,939	425,066
33.5	9.6	2490	351	137,344	12,836	6,717,721	469,771
38.1	10.6	2791	388	153,207	14,186	7,476,148	519,177
43.3	11.7	3127	428	170,888	15,678	8,319,799	573,779
49.2	12.9	3503	473	190,593	17,327	9,258,212	634,124
55.8	14.3	3924	523	212,553	19,149	10,301,989	700,816
63.2	15.8	4395	578	237,023	21,163	11,462,910	774,521
71.6	17.5	4921	639	264,290	23,389	12,754,073	855,978
81.1	19.3	5509	706	294,671	25,848	14,190,031	946,002
91.7	21.3	6166	781	328,518	28,567	15,786,958	1,045,494
104	23.6	6901	863	366,226	31,571	17,562,832	1,155,449
117	26.0	7722	953	408,231	34,892	19,537,628	1,276,969
132	28.8	8640	1054	455,021	38,561	21,733,546	1,411,269
150	31.8	9665	1164	507,138	42,617	24,175,252	1,559,694
169	35.2	10,810	1287	565,184	47,099	26,890,158	1,723,728

10. Open the spreadsheet "Activity 5B_cdc13MutationRate" at https://hackettmolecularbiology.blogspot.com/ and make a copy of the spreadsheet that you can edit. The data have been entered in the top section of the spreadsheet. Underneath this top section are separate sections used to calculate the mutation rate for each condition. At the far right side of one of these mutation rate calculation sections, find the turquoise-colored box containing the median r_0 value for the median number of new mutant cells. Look up this r_0 value in the table above and estimate the value of m, the mean number of mutations per culture. Type this m value into the magenta-colored m box in the spreadsheet.

The value for m determined from the median r_0 value is not ideal because it does not take into account the actual number of new mutant cells, r, present in each of the individual cultures. A better estimate of m is obtained by evaluating the following equations using several different adjacent values for m. The value for m that makes the final sum for all cultures equal to zero is the best estimate of m. (Note that the initial estimate of m from r_0 is necessary because many different values of m can make the sum evaluate to be zero. The initial estimate of m from r_0 places the initial estimate for m close to the best final value for m).

$$x = \frac{11.6}{r/m - \ln m + 4.5} - 2.02$$

$$d = (1 - 4.5 + \ln m)/11.6$$

$$\sum_{all\ cultures} [x(x + 2.02)^2 d + x(x + 2.02) - 2(x + 2.02)d - 1] = 0$$

The value for m determined from the above Maximal Likelihood Method is then used to calculate the mutation rate by dividing m by the average number of cell divisions (or average number of new cells) in the growth of the cultures. This gives the mutation rate in mutations/cell division.

11. So far, the estimate of the mean number of mutations m in the magenta box in your spreadsheet is based only on the median culture. The spreadsheet calculates values for x and d for each culture. It also calculates the total sum for all cultures and shows this value in the yellow box. The current value in the yellow box is probably not zero. Now you will refine the estimate of m using the data from all of the cultures. The best estimate of m is obtained when the total sum in the yellow box is equal to zero. Slightly change the m value in the magenta box up or down until the value in the yellow box is very close to zero. At this point, the calculated mutation rate in the green box is the most correct.

12. Repeat the above steps to calculate a mutation rate for each condition.

13. The spreadsheet has a "Mutation Rate" sheet, a "Linear Graph" sheet, and a "Log Graph" sheet that summarize the mutation rates from the "Data and Calculations" sheet. Notice that the "Mutation Rate" sheet includes the mutation rates that you calculated for the rates of all deletions that extend any distance past *CAN1* and the rate of deletions that do not extend past *LYS5*. This sheet also calculates the rate of deletions that do extend past *LYS5* by subtracting the other two rates. The graph included at the end of this activity plots the mutation rates obtained from these data. If your mutation rates differ significantly from these, repeat the above steps to calculate mutation rate. You can record your calculated mutation rates here for your reference:

	wildtype 22°C	wildtype 37°C	cdc13-1 22°C	cdc13-1 37°C
ade5Δ can1Δ +/-lys5 mutation rate (deletion extends any distance past *CAN1*)				
ade5Δ can1Δ lys5Δ mutation rate (deletion extends further than *LYS5*)				
ade5Δ can1Δ +LYS5 mutation rate (deletion does not pass *LYS5*)				

14. Consider the map of chromosome VII:
Are the rates of deletions extending into different regions of the chromosome proportional to the lengths of those regions of the chromosome? Or is there an enrichment for deletions in particular regions? Is there a difference in different conditions?

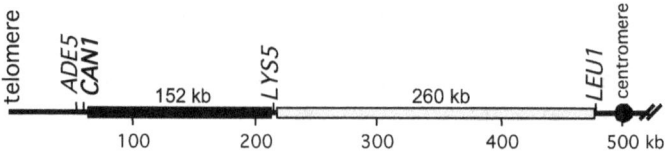

116 | Unit 5   Class Activity 5B: Telomeres and Mutation Rate

PART 3: Discussion Questions

14. Do the data and calculations make sense in the context of the underlying biology? Explain.

15. Explain how this experiment provides evidence that telomeres protect the ends of chromosomes and help to maintain genomic stability. Refer to specific proteins, genes, and data/calculations.

16. Why does it make sense that heat-shocked *cdc13-1* cells are prone to cell cycle arrest?

17. Consider how the proportion of deletions that did not extend past *LYS5* compared to the proportion of deletions that did extend past *LYS5* differed in the heat-shocked *cdc13-1* yeast compared to the other conditions. Why does this make sense given the kind of DNA damage you would expect to occur more frequently in heat-shocked *cdc13-1* cells than in wildtype cells?

18. Is the number of colonies on a canavanine plate more directly related to mutation rate (*can1* mutations/cell division) or to mutation frequency (*can1* mutant cells/all cells)? Explain.

19. Why did we need to use many independent liquid cultures for each experimental condition?

20. Briefly explain the general roles of mutation and selection in cancer.

21. In relation to the *CAN1* gene, explain the causes of mutation and selection in this experiment.

22. Describe some benefits and drawbacks of using yeast as a model system to study phenomena related to human cancer.

Solution to Mutation Rate Calculations:

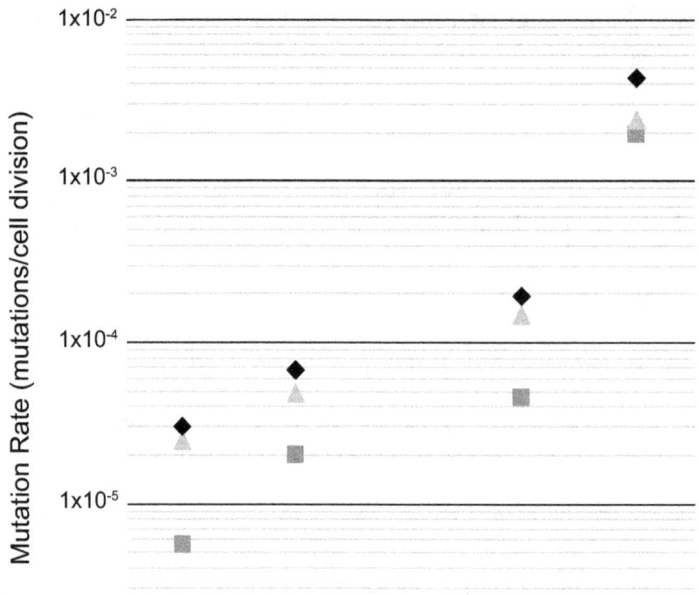

	wildtype 22°C	wildtype 37°C		cdc13-1 22°C	cdc13-1 37°C
◆ ade5Δ can1Δ +/-lys5 mutation rate (deletion extends any distance past *CAN1*)	2.99x10⁻⁵	6.81x10⁻⁵		1.92x10⁻⁴	4.36x10⁻³
▲ ade5Δ can1Δ lys5Δ mutation rate (deletion extends further than *LYS5*)	2.44x10⁻⁵	4.82x10⁻⁵		1.47x10⁻⁴	2.40x10⁻³
■ ade5Δ can1Δ +LYS5 mutation rate (deletion does not pass *LYS5*)	5.52x10⁻⁶	1.98x10⁻⁵		4.53x10⁻⁵	1.95x10⁻³

Chapter 5 Reading #5 and Discussion Questions

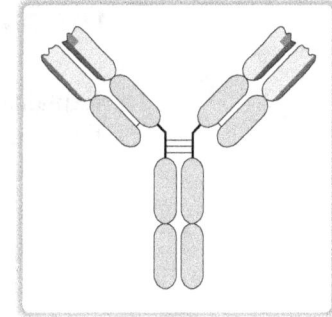

Reading: *Molecular Biology: Concepts for Inquiry*
Chapter 5: Section 5.3

Online Quiz: 5.3

Discussion Questions:

1. Given the following data, explain where the estimate of 3.5 million antibodies based on combinations of VDJ segments comes from. Assume homologous alleles are identical. Use the following data to calculate an estimate of the number of different antibody genes that can be assembled.
   There are two different genes for the light chain, either one or the other is productively rearranged:
   Kappa light chain locus: 40 V and 5 J segments
   Lambda light chain locus: 30 V and 4 J segments
   There is one gene for the heavy chain:
   Heavy chain locus: 65 V, 27 D, and 6 J segments

2. What other two processes add to the diversity of antibodies? Explain.

3. Explain why it makes sense that memory B cells have membrane-bound antibodies.

4. How are both negative and positive selection of antibodies key processes in the immune system?

5. How is the "genome evolution" in the immune system similar to and different from that observed in cancer and in the evolution of species?

6. An allergy occurs when the immune system reacts to something harmless from the environment as if it were a dangerous pathogen. In severe allergic reactions, the immune system can overreact so much that death can occur. The science is not yet well-understood, but there is increasing evidence that exposures to potential allergens in infancy and early childhood can sometimes help to reduce the risk of developing allergies. For example, early exposures to food allergens like peanuts and exposure to nonsterile environments like barnyards are associated with a lower incidence of allergies. What does this suggest about how "training" of immune cells might occur in young children?

Unit 5 Self-Assessment

1. Explain, in terms of cell cycle regulation, why *p53* is more commonly mutated in malignant tumors than most other tumor suppressor genes or oncogenes.

2. Why is it thought that telomerase is expressed in germline cells (cells that produce eggs/sperm), but not in most somatic cells?

3. What is generally different about childhood cancers and adult cancers in terms of how the cancer arises (consider cell type, timing, number of mutant genes). In other words, why don't adults usually get childhood cancers and why don't children usually get adult cancers?

4. How does a high mutation rate only at antibody genes in (a) pre-B cells and (b) mature B cells contribute to antibody diversity. Include a brief description of how mutations occur in each process. Include the roles of positive or negative selection in each process.

5. Species #1 has 30 chromosomes, species #2 has 30 chromosomes, and species #3 has 32 chromosomes. The number of blocks of conserved synteny when considering pairs of species are:
    #1 and #2: 200 blocks
    #1 and #3: 40 blocks
    #2 and #3: 100 blocks
   a. Which 2 species are likely to be most closely related to each other?
   b. Explain your reasoning.

6. Explain how gene duplication can facilitate the evolution of a new gene. Include the roles of mutation and selection in your answer.

7. How does a cell tell the difference between a telomere and a DNA break?

8. Write "T" for "tumor-suppressor gene" or "O" for "oncogene" for each of the following statements:
   a. Normally <u>limits</u> cell division.
   b. Normally <u>promotes</u> cell division.
   c. Is more likely to be "off" in <u>normal</u> cells.
   d. Is more likely to be "off" in <u>tumor</u> cells.
   e. More likely to have a mutation of <u>one</u> copy of the gene in tumors.
   f. More likely to have mutations of <u>two</u> copies of the gene in tumors.
   g. More likely to have <u>loss</u> of function mutations in tumors.
   h. More likely to have <u>gain</u> of function mutations in tumors.

9. Explain the following about the mutation rate lab experiment we studied:
   a. Explain at the molecular level why the heat-shocked *cdc13-1* yeast had a higher rate of deletion of the *CAN1* and *ADE5* genes than wildtype yeast or *cdc13-1* yeast that hadn't been heat-shocked.

   b. How did we know that the elevated mutation rate in heat-shocked *cdc13-1* yeast resulted from elevated rates of large deletions, but not elevated rates of point mutations?

10. Given the following pedigree for an extremely rare disease where affected individuals are represented by filled shapes:

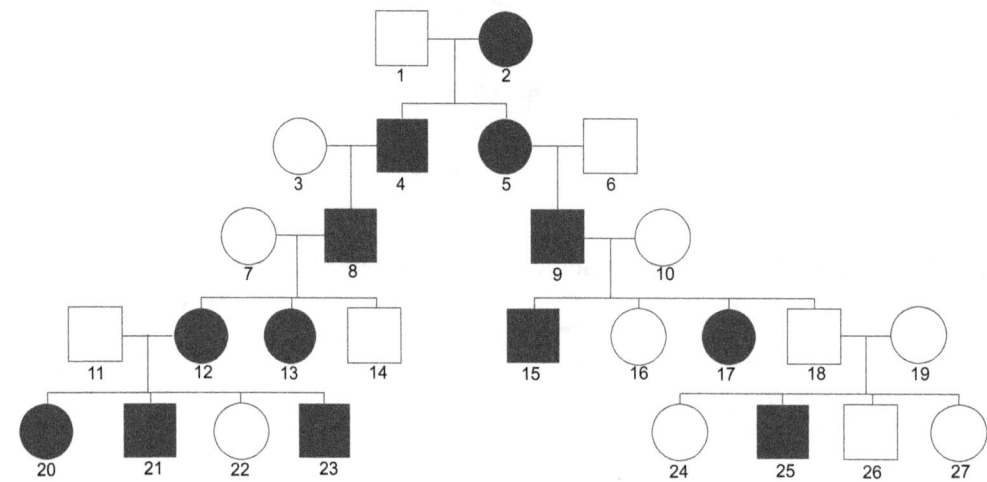

   a. Considering ONLY the information provided in the pedigree, is the mode of inheritance more likely to be dominant or recessive? Explain why.

   b. With the additional information that the disease is a rarely-fatal childhood cancer, would you need to reevaluate your answer to (a). Explain why or why not with a thorough explanation of genetics.

   c. Is the gene studied in this pedigree more likely to be a tumor suppressor gene or an oncogene?

   d. Explain the genetics of how it was possible for person #18 to be unaffected.

Unit 6

Emerging Molecular Biology, Biotechnology and Medicine

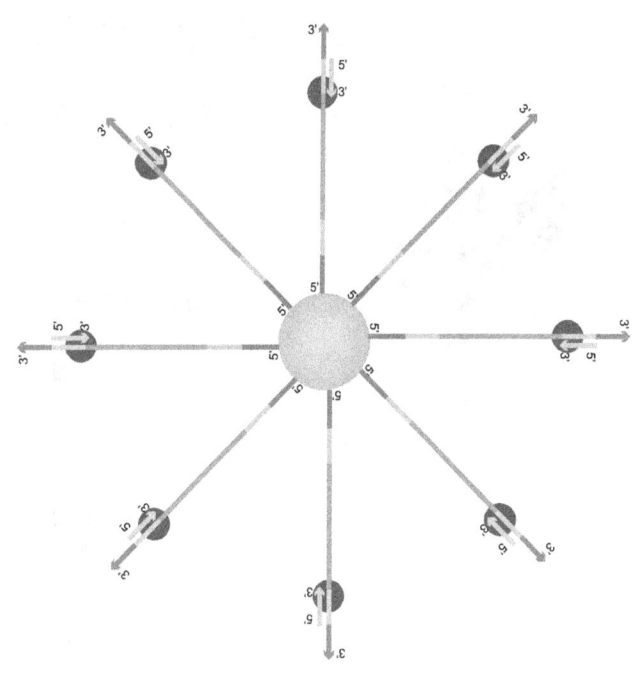

The previous units have focused on the fundamental concepts that underlie molecular biology and biotechnology. This unit will focus on how these fundamental concepts are currently being applied to improve our ability to understand and treat diseases, with a focus on recently developed and emerging technologies. Improved technologies are beginning to bring precision genetics-based medicine to the clinic more frequently.

To achieve the deepest level of understanding, you are encouraged to complete the class activities, textbook readings and question sets in the order listed below.

    Suggested order:
        Chapter 6 Reading #1 and Discussion Questions
        Class Activity 6A: Next Generation DNA Sequencing
        Chapter 6 Reading #2 and Discussion Questions
        Chapter 6 Reading #3 and Discussion Questions
        Class Activity 6B: Curing a Genetic Disease with CRISPR/Cas9
        Unit 6 Self-Assessment Questions

    Prior knowledge (review if necessary):
- Structure of DNA, RNA, and proteins (in Chapters 1, 3)
- Complementary DNA nucleotides (in Chapters 1, 3)
- Enzyme structure and function (in Chapter 2)
- DNA replication (in Chapter 3)
- PCR (in Chapter 3)
- DNA sequencing (in Chapter 3)
- Stem cells (in Chapter 4)
- Organization of a gene (in Chapters 2, 4)

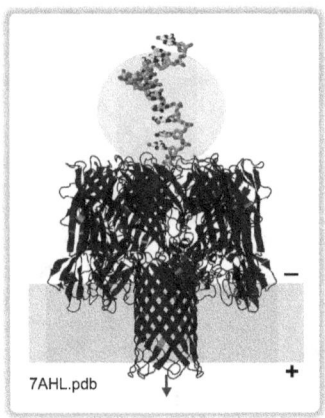

Chapter 6 Reading #1 and Discussion Questions

Reading: *Molecular Biology: Concepts for Inquiry*
Chapter 6: Introduction and Section 6.1

Online Quiz: 6.1

Discussion Questions:

1. What were the general shifts in sequencing strategy that drove the development of next-generation sequencing techniques?

2. How do the capabilities of next-generation sequencing differ from Sanger sequencing?

3. Why is data storage a significant concern in next-generation sequencing? How are some techniques better than others in this regard?

4. In what circumstances would you still choose to do Sanger sequencing?

5. If you needed to sequence DNA, under what circumstances would you choose to use a long-read next generation sequencing strategy versus a short-read next generation sequencing strategy.

6. How did the strategy used in the Human Genome Project provide extra checks on the accuracy of the assembly in ways that would not have been similarly possible if next generation sequencing had been used to sequence the genome for the first time?

7. How do you think the reduction in cost of next generation sequencing might lead to more personalized medicine?

Class Activity 6A: Next Generation DNA Sequencing

A selected set of Next Generation DNA Sequencing techniques:

    454 Sequencing
    Illumina Sequencing
    Semiconductor Sequencing (Ion Torrent)
    Pacific Biosystems real time single molecule sequencing
    Nanopore Sequencing

1. As a group, you should start by learning about 454 sequencing. First, familiarize yourself with the information in Section 6.1 of the textbook. Ask each other clarifying questions as you analyze Figures 6.3 – 6.6 in detail.

2. Then split up and each person should become an expert at understanding one of the additional sequencing techniques listed above. Your textbook is a good place to start. Also visit the websites for the companies and look for additional explanations there. Write down pros and cons for your technique. Also find a visual presentation of the science behind the technique that can be accompanied by a detailed explanation from you. The visual component could be a web animation or a textbook figure, but you will need to pause any animation frequently and explain it to your groupmates and answer their questions.

   After doing your research, present to your group:
   a. The science behind the sequencing technology (how it works)
   b. The pros
   c. The cons

3. Assume that your group members are scientists trying to decide which next generation sequencing machine to buy within the next year. Consider the information you discussed above as well as any other information in Table 6.1 of the textbook. Which machine would you choose and why if your goal is to do the following (the ideal answer for each scenario may be different and there may be more than one reasonable choice for each):
   a. Sequence the genomes of microorganisms whose genomes have never been sequenced before.
   b. Sequence many human cancer genomes.
   c. Sequence the exons from patients with rare genetic diseases.
   d. Start a company to make the sequencing of personal genomes accessible to physicians and/or the general public.

Chapter 6 Reading #2 and Discussion Questions

Reading: *Molecular Biology: Concepts for Inquiry*
Chapter 6: Sections 6.2A-B

Online Quiz: 6.2A-B

Discussion Questions:

1. What is the difference between a cell and a virus? Can a virus infect any cell? Explain.

2. What is a retrovirus? In a previous chapter, you learned how one retrovirus protein is used in genetic engineering. Which one and for what purpose is it used?

3. Why is there usually a lag period between the time that a person is infected with HIV and the time that they develop AIDS? Why are drug cocktails used to treat HIV?

4. Why can't HIV be cured by the body's immune system? In what fundamental way do you think an HIV infection (doesn't go away) differs from an infection by a virus like influenza that does go away after an illness?

5. How do adenoviruses differ from retroviruses? What are the pros and cons of using each type of virus for gene therapy? How do AAV viruses differ from lentiviruses? What are the pros and cons of using each type of virus for gene therapy?

6. In what ways has a better understanding of the biology of viruses as well as a better understanding of the regulation of expression in the human genome led to the development of safer, more effective gene therapy vectors?

7. Why do you think gene therapy technology has developed more slowly than next generation sequencing technology?

Chapter 6 Reading #3 and Discussion Questions

Reading: *Molecular Biology: Concepts for Inquiry*
Chapter 6: Sections 6.2C-D

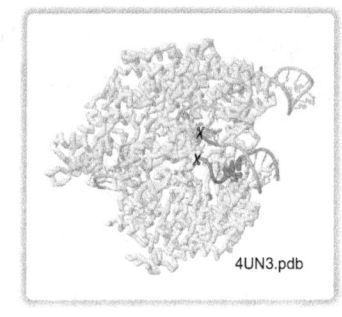

Online Quiz: 6.2C-D

Discussion Questions:

1. Compare and contrast the following gene editing strategies: zinc finger endonucleases, TALENs, CRISPR/Cas9.

2. In what type of situation do you think it would be useful to be able to use ZFNs/TALENs/CRISPRs clinically? Is this different from the type of situation in which you would want to use traditional gene therapy?

3. Explain the concern about off-target effects and zinc finger endonucleases.

4. For each of the following situations, what kind of gene therapy would you want to try to use to treat/cure the disease?
    a. Hemophilia (No blood cells make a protein required for blood clotting)
    b. Leukemia (Some bone marrow cells are cancerous and crowd out healthy bone marrow cells).
    c. Marfan syndrome (A dominant negative mutation causes defects in connective tissue elasticity and the increased release of a growth factor in many organs. The aorta can burst, eye lenses can detach, joints are extra flexible).

5. How are antibody "biological drugs" made?

6. Which example of novel treatment strategies described in section 6.2D did you find most interesting and why?

Class Activity 6B: Curing a Genetic Disease with CRISPR/Cas9

Work with others in your small class group to complete these questions. Thoroughly discuss one question at a time without skipping ahead. Use your logic skills to answer the questions - do not look up answers in your textbook or in another reference.

Your goal in this activity is to design a CRISPR guide RNA molecule that could be used to cure a genetic disease. The DNA sequence included below is a portion of the *CFTR* gene that is mutated in cystic fibrosis, but you could extend the principles in this activity to any genetic disease where the mutation is known.

1. Review how CRISPR/Cas9 works using the diagram below. Sources for further review are the textbook and this interactive simulation at HHMI Biointeractive: http://media.hhmi.org/biointeractive/click/CRISPR/

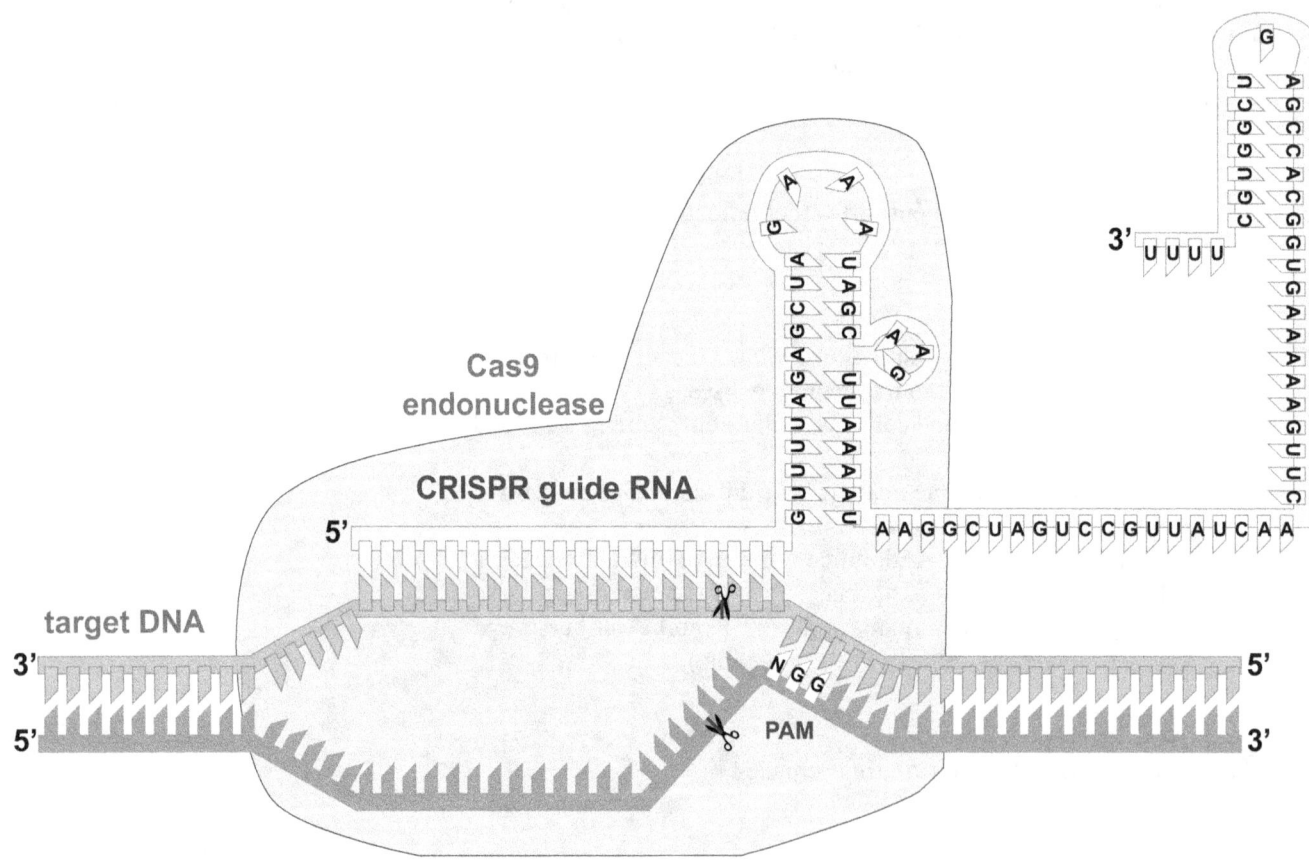

2. Review the biology of cystic fibrosis. The NIH Genetics Home Reference is a great resource for learning about genetic diseases. https://ghr.nlm.nih.gov/condition/cystic-fibrosis

    a. What is the mode of inheritance of cystic fibrosis? (Autosomal dominant, autosomal recessive, sex-linked dominant, sex-linked recessive, something more complicated)?

    b. What kind of mutation and how many alleles must be mutated to cause the disease? (Two-allele recessive loss of function; or one-allele dominant haploinsufficiency, gain of function, or dominant negative)?

    c. Explain what kind of change to the DNA would be necessary to cure the patient. One or both alleles? Would you want to use NHEJ or homologous recombination for repair and why?

Part of the genomic DNA sequence of the *CFTR* gene where the top strand is the 5' to 3' sense strand:
  Dark highlight: exon #11 of the 27 exons in *CFTR*; translation is shown.
  White highlight: location of the 3 nucleotides commonly deleted in cystic fibrosis (ΔCTT).
  Black box: Phenylalanine 508. The 3 nucleotides that encode F508, TTT, are directly below the box.
  No highlight: intron sequence
  The full sequence file can be downloaded from https://hackettmolecularbiology.blogspot.com/.

```
                                                                                                80900
5'
ACTTGGCAAC TGTTAGCTGT TACTAACCTT TCCCATTCTT CCTCCAAACC TATTCCAACT ATCTGAATCA TGTGCCCCTT CTCTGTGAAC CTCTATCATA
TGAACCGTTG ACAATCGACA ATGATTGGAA AGGGTAAGAA GGAGGTTTGG ATAAGGTTGA TAGACTTAGT ACACGGGGAA GAGACACTTG GAGATAGTAT
3'
                                                                                                81000
ATACTTGTCA CACTGTATTG TAATTGTCTC TTTTACTTTC CCTTGTATCT TTTGTGCATA GCAGAGTACC TGAAACAGGA AGTATTTTAA ATATTTTGAA
TATGAACAGT GTGACATAAC ATTAACAGAG AAAATGAAAG GGAACATAGA AAACACGTAT CGTCTCATGG ACTTTGTCCT TCATAAAATT TATAAAACTT
                                                                                                81100
TCAAATGAGT TAATAGAATC TTTACAAATA AGAATATACA CTTCTGCTTA GGATGATAAT TGGAGGCAAG TGAATCCTGA GCGTGATTTG ATAATGACCT
AGTTTACTCA ATTATCTTAG AAATGTTTAT TCTTATATGT GAAGACGAAT CCTACTATTA ACCTCCGTTC ACTTAGGACT CGCACTAAAC TATTACTGGA
                                                                                                81200
                        T  S   L  L  M   V  I  M   G  E  L   E  P  S   E  G  K   I  K  H   S  G  R   I  S
AATAATGATG GGTTTTATTT CCAGACTTCA CTTCTAATGG TGATTATGGG AGAACTGGAG CCTTCAGAGG GTAAAATTAA GCACAGTGGA AGAATTTCAT
TTATTACTAC CCAAAATAAA GGTCTGAAGT GAAGATTACC ACTAATACCC TCTTGACCTC GGAAGTCTCC CATTTTAATT CGTGTCACCT TCTTAAAGTA
                                                                                                81300
 F  C  S   Q  F  S   W  I  M   P  G  T   I  K  E   N  I  I [F] G  V   S  Y  D   E  Y  R   Y  R  S   V  I  K  A
TCTGTTCTCA GTTTTCCTGG ATTATGCCTG GCACCATTAA AGAAAATATC ATCTTTGGTG TTTCCTATGA TGAATATAGA TACAGAAGCG TCATCAAAGC
AGACAAGAGT CAAAAGGACC TAATACGGAC CGTGGTAATT TCTTTTATAG TAGAAACCAC AAAGGATACT ACTTATATCT ATGTCTTCGC AGTAGTTTCG
                                                                                                81400
 C  Q  L   E  E
ATGCCAACTA GAAGAGGTAA GAAACTATGT GAAAACTTTT TGATTATGCA TATGAACCCT TCACACTACC CAAATTATAT ATTTGGCTCC ATATTCAATC
TACGGTTGAT CTTCTCCATT CTTTGATACA CTTTTGAAAA ACTAATACGT ATACTTGGGA AGTGTGATGG GTTAATATA TAAACCGAGG TATAAGTTAG
                                                                                                81500
GGTTAGTCTA CATATATTTA TGTTTCCTCT ATGGGTAAGC TACTGTGAAT GGATCAATTA ATAAAACACA TGACCTATGC TTTAAGAAGC TTGCAAACAC
CCAATCAGAT GTATATAAAT ACAAAGGAGA TACCCATTCG ATGACACTTA CCTAGTTAAT TATTTTGTGT ACTGGATACG AAATTCTTCG AACGTTTGTG
                                                                                                81600
                                                                                                       3'
ATGAAATAAA TGCAATTTAT TTTTTAAATA ATGGGTTCAT TTGATCACAA TAAATGCATT TTATGAAATG GTGAGAATTT TGTTCACTCA TTAGTGAGAC
TACTTTATTT ACGTTAAATA AAAAATTTAT TACCCAAGTA AACTAGTGTT ATTTACGTAA AATACTTTAC CACTCTTAAA ACAAGTGAGT AATCACTCTG
                                                                                                       5'
```

3. A mutation that commonly causes cystic fibrosis is the deletion of the nucleotides CTT that result in the deletion of phenylalanine 508 (ΔF508) in the *CFTR* gene. The relevant portion of the normal *CFTR* DNA sequence is shown above and the locations of the deletion and the F508 codon are marked. How does the deletion of CTT cause ΔF508 without affecting other amino acids?

4. Design the sequence of the 20-nucleotide "spacer" region of a CRISPR guide RNA. This is the region of the CRISPR RNA that is complementary to the sequence in the target genomic DNA. Refer to the diagram on the previous page for how the CRISPR RNA/target DNA will need to be positioned relative to the PAM sequence in the DNA. Pay attention to where the 5' and 3' ends of nucleic acids are positioned in the two figures.

Note: The first nucleotide in the CRISPR guide RNA is often designed to be a G because the U6 promoter that is often used to express artificial CRISPR RNAs has a preference for starting transcription at G. However, this is not a requirement of the Cas9 enzyme itself. The guide RNA will work with one extra nucleotide on the 5' end, so if your sequence doesn't start with G, just plan to add an extra 5' G.

Class Activity 6B: Curing a Genetic Disease with CRISPR/Cas9

5. Check the sequence you designed:
    a. Is it 20 nucleotides long (excluding any extra G that you tacked onto the 5' end)?
    b. When written 5' to 3', are the next 3 nucleotides in the DNA after the 3' end NGG? Cas9 needs the NGG PAM sequence to be there in the DNA for it to be able to find the target DNA, but the PAM sequence should not be part of the guide RNA.
    c. When written 5' to 3', is it identical to one of the strands of DNA in the 5' to 3' direction except that the RNA contains Us instead of Ts? And is the PAM on that DNA strand?
    d. Do the 20 nucleotides you chose include the ΔCTT nucleotides (or complement)? If so, oh no! This design won't work because this wildtype sequence isn't in the patient's DNA. Please try again.
    e. Will Cas9 cut close enough to the mutation site that you'll be able to design a patch that restores the wildtype sequence? Ideally the change you plan to make should be within 10bp of the cut site and definitely not more than 100bp away.

6. Design a DNA sequence to act as a patch to repair the patient's gene through homologous recombination. It should work if you include 50-80bp of homologous sequence on either side of the change. Consider whether there could be a way to design the patch DNA so that it would code for the wildtype protein, but couldn't be re-cut by the CRISPR/Cas9 that you designed above.

7. Describe the strategies you would need to use to deliver your system into the patient's cells. What would you want to do to test your system before actually trying it in a patient. After trying it in a patient, how would you know if it worked? What kind of safety controls would you try to build in?

8. A worry when actually designing CRISPR RNA molecules is that there might be off-target effects, meaning CRISPR/Cas9 might cut someplace that you didn't intend. What could you do to minimize the chance of off-target effects?

9. Compare your design and reasoning to the designs published in *Cell Stem Cell* 13 (2013) 653–658. A guide RNA that targets a sequence in exon 11 is shown in the top part of Figure 1F and a guide RNA that targets intron 11 at about 81480bp is shown in the bottom part of Figure 1F.
   http://dx.doi.org/10.1016/j.stem.2013.11.002

10. What do you think are the implications of this system being simple enough that a novice can create a design for editing a genome?

Design reference - Zhang Lab CRISPR page: https://zlab.bio/guide-design-resources

Unit 6 Self-Assessment

1. Explain why the choice of which next-generation sequencing technique to use might differ when sequencing a species' genome for the first time versus resequencing an individual's genome for a species that has already been sequenced.

2. What is the main reason that Sanger sequencing is unlikely to become obsolete in the near future?

3. Give two different ways that genome editing might be used to try to cure HIV. Which would you prefer and why?

4. Explain two major problems with early gene therapy. How were these problems overcome?

5. Explain why TALENs are easier to use for genome editing than zinc finger endonucleases. Explain why CRISPR/Cas9 is easier to use for genome editing than TALENs.

6. If you were planning to use genome editing to cure a person's genetic disease, how would your strategy differ depending on whether the person had a recessive disease or a dominant disease? If it were a dominant disease, how would the mechanism of dominance (haploinsufficiency, gain of function, dominant negative) affect your strategy?

PDB References

PDB files (noted with "PDB ID") were downloaded from www.rcsb.org, H.M. Berman, J. Westbrook, Z. Feng, G. Gilliland, T.N. Bhat, H. Weissig, I.N. Shindyalov, P.E. Bourne. (2000) The Protein Data Bank Nucleic Acids Research, 28: 235-242.

Jennifer Hackett used Jmol software to format images based on PDB files.

Cover:

*Eco*RI, PDB ID: 1ERI, Kim, Y.C., *et al*. Refinement of Eco RI endonuclease crystal structure: a revised protein chain tracing. *Science* 249, 1307-1309 (1990).

Unit 1:

ADP-glucose pyrophosphorylase, PDB ID: 1YP2, Jin, X., *et al*. Crystal structure of potato tuber ADP-glucose pyrophosphorylase. *EMBO J.* 24, 694-704 (2005).

Glycogen phosphorylase, PDB ID: 8GPB, Barford, D., Hu, S.H., & Johnson, L.N. Structural mechanism for glycogen phosphorylase control by phosphorylation and AMP. *J. Mol. Biol.* 218, 233-260 (1991).

Hexokinase, PDB ID: 1HKG, Steitz, T.A., Shoham, M., & Bennett Jr., W.S. Structural dynamics of yeast hexokinase during catalysis. *Philos.Trans.R.Soc.London,Ser.B* 293, 43-52 (1981).

Phosphoglucomutase, PDB ID: 3PMG, Liu, Y., Ray, W.J., & Baranidharan, S. Structure of rabbit muscle phosphoglucomutase refined at 2.4 A resolution. *Acta Crystallogr.,Sect.D* 53, 392-405 (1997).

RNA chaperone, PDB ID: 3MW6, Chaulk, S., *et al*. N. meningitidis 1681 is a member of the FinO family of RNA chaperones. *RNA Biol.* 7, 112-119 (2010).

tRNA, PDB ID: 2QNH, Korostelev, A., *et al*. Interactions and dynamics of the Shine Dalgarno helix in the 70S ribosome. *Proc. Natl. Acad. Sci. USA* 104, 16840-16843 (2007).

Unit 2:

Abl kinase, inactive with imatinib, PDB ID: 1IEP, Nagar, B., *et al*. Crystal structures of the kinase domain of c-Abl in complex with the small molecule inhibitors PD173955 and imatinib (STI-571). *Cancer Res.* 62, 4236-4243 (2002).

Antibody, PDB ID: 1IGT, Harris, L.J., *et al*. Refined structure of an intact IgG2a monoclonal antibody. *Biochemistry* 36, 1581-1597 (1997).

Collagen, PDB ID: 1CAG, Bella, J., Eaton, M., Brodsky, B., Berman, H.M. Crystal and molecular structure of a collagen-like peptide at 1.9 A resolution. *Science* 266: 75-81 (1994).

GFP, PDB ID: 1GFL, Yang, F., Moss, L.G., & Phillips Jr., G.N. The molecular structure of green fluorescent protein. *Nature Biotechnology* 14, 1246-1251 (1996).

Kap95p:RanGTP, PDB ID: 2BKU, Lee, S.J., *et al*. Structural Basis for Nuclear Import Complex Dissociation by Rangtp. *Nature* 435, 693 (2005).

RFP, PDB ID: 1GGX, Wall, M.A., Socolich, M., & Ranganathan, R. The structural basis for red fluorescence in the tetrameric GFP homolog DsRed. *Nat. Struct. Mol. Biol.* 7, 1133-1138 (2000).

Ribosomal subunit, large, PDB ID: 2WDL, Voorhees, R.M., *et al*. Insights into substrate stabilization from snapshots of the peptidyl transferase center of the intact 70S ribosome. *Nat. Struct. Mol. Biol.* 16, 528-533 (2009).

Ribosomal subunit, small, with tRNAs and mRNA, PDB ID: 2WDK, Voorhees, R.M., *et al*. Insights into substrate stabilization from snapshots of the peptidyl transferase center of the intact 70S ribosome. *Nat. Struct. Mol. Biol.* 16, 528-533 (2009).

Unit 3:

DNA, PDB ID: 1BNA, Drew, H.R., *et al*. Structure of a B-DNA dodecamer: conformation and dynamics. *Proc. Natl. Acad. Sci. USA* 78, 2179-2183 (1981).

Unit 4:

Nucleosome, PDB ID: 1AOI, Luger, K., *et al*. Crystal structure of the nucleosome core particle at 2.8 A resolution. *Nature* 389, 251-260 (1997).

Unit 6:

CRISPR/Cas9 binding target DNA, PDB ID: 4UN3, Bae, B., *et al*. Structural Basis of Pam-Dependent Target DNA Recognition by the Cas9 Endonuclease. *Nature* 513, 569 (2014).

Appendices:

RNA polymerase initiating transcription, PDB ID: 4XLN, Anders, C., *et al*. Structure of a bacterial RNA polymerase holoenzyme open promoter complex. *Elife* 4, e08504 (2015).

TrpR, PDB ID: 1TRR, Lawson, C. L. & Carey, J. Tandem binding in crystals of a trp repressor/operator half-site complex. *Nature*, 366,178-182 (1993).

Appendix 1: Answers to Self-Assessments

Unit 1

1. a) C=O double bond, C-O single bond, O-H bond. b) δ^- :both oxygens, δ^+ :H and the C between the Os. c) 8. d) The carboxylic acid group behaves as an acid because it donates a hydrogen ion.
2. The sidechains should be nonpolar and hydrophobic because if they were polar and hydrophilic they would be attracted to the water outside of the membrane.
3. a) 13. b) carboxylic acid, amino. c) amino acid. d) yes.
4. row 1: glycerol, fatty acids; row 2: protein; row 3: monosaccharide or sugar; row 4: DNA and RNA (nucleic acid is incorrect because that also describes nucleotides).
5. The structures of starch and cellulose differ because the bonds that link glucose monomers to each other are in different positions in starch and cellulose. The bond angles in cellulose orient the hydroxyl groups such that they can hydrogen bond with another polymer of cellulose. A cellulose polymer hydrogen bonds with other cellulose polymers better than it hydrogen bonds with water, so cellulose does not dissolve in water. The bond angles in a starch polymer do not allow it to hydrogen bond very well with other starch polymers. Therefore, the hydroxyl groups in starch are able to hydrogen bond with water, so starch dissolves in water.
6. Water is a polar molecule and it forms hydrogen bonds with other water molecules. The only intermolecular forces (IMFs) between oil molecules are dispersion forces, so oil is not attracted to the water. The fact that all of the water molecules are strongly attracted to each other through H-bonds allows the water to form a separate layer, leaving the oil in its own layer. (The answer should not imply that there is a repulsive force between oil and water).
7. G-C base pair because it has 3 H-bonds.
8. [structure of phosphate ion PO_4^{3-}]
9. Missing reactants: 2 water molecules.
10. [structure showing amide + H_2O]
11. Breaking bonds takes energy. Making bonds releases energy.
12. a) ΔH is negative because heat is released when more IMFs form. ΔS is negative because the structure of a folded protein is more ordered than an unfolded protein (it can adopt fewer shapes). The sign of ΔG depends on temperature, but because protein folding is spontaneous, the sign of ΔG must be negative when a protein folds. b) When the temperature is high the sign of ΔG is positive and the protein does not fold.
13. Yes, the actual ΔG could be negative and the reaction could therefore be spontaneous if the concentrations of the reactants are high and the concentration of the product is low. (See the equation that calculates an actual ΔG from a standard $\Delta G°$ and the concentrations).
14. a) Coupled reactions tend to proceed through an alternate reaction pathway where intermediate reactions (that are spontaneous individually) share products and reactants. Enzymes preferentially speed up reactions in the coupled reaction pathway. The sum of the ΔGs for the coupled reactions is negative and is equal to the sum of the ΔGs for the reactions when they take the alternate reaction pathway. b) To make ATP, which is nonspontaneous. In the absence of oxygen, which is required for cellular respiration, fermentation couples the nonspontaneous synthesis of ATP to the spontaneous production of ethanol and carbon dioxide from glucose.
15. Organelles allow eukaryotes to concentrate reactants in and export products from specific regions of the cell, thus increasing the likelihood that reactions will be spontaneous (see the equation that calculates an actual ΔG from a standard $\Delta G°$ and the concentrations). Prokaryotes are much smaller than eukaryotic cells and concentrations of reactants are already high enough in the smaller prokaryotic cells that they do not need to use organelles to allow for spontaneity.
16. Humans only produce enzymes that digest particular glucose polymers like starch and glycogen. We do not produce enzymes that digest cellulose.
17. a) rough ER. b) lysosome. c) mitochondria. d) chloroplast.
18. a) False. b) False. c) False. d) True. e) False.
19. Chloroplast, mitochondria, nucleus, lysosome, rough ER, smooth ER.
20. Products. The 6 carbon dioxides and 6 waters together contain more stable bonds than the glucose and 6 oxygen molecules together. More energy is released when the products form from the transition state than was used to break the bonds in the reactants. This is true for any combustion reaction since combustion reactions are exothermic.

Unit 2

1. a) Dominant because only one allele is mutant in patients with the disease. b) Dominant negative because the mutant proteins interfere with the normal function of the wildtype proteins.
2. Outside.
3. a) AUGCCCAGCCUGGGAUGACGA. b) Met-Pro-Ser-Leu-Gly.
4. Any two: remove the introns, add a ribosome binding site, replace the eukaryotic promoter with a bacterial promoter, replace the eukaryotic 3'UTR with a sequence containing a bacterial transcriptional terminator.
5. a) Gene B because the nonsense mutation in the affected family members creates a very premature stop codon, thus certainly preventing the production of a functional protein. The silent mutation in gene A in the affected family members encodes the same amino acid (histidine), so this mutation is very unlikely to affect the function of the gene. b) They should try to add more wildtype protein because the premature stop codon creates a loss of function for that allele. c) The disease gene most likely encodes a protein that is not an enzyme. Since the disease phenotype is likely caused by a dominant haploinsufficiency, the normal allele does not produce a sufficient dose of the protein to create a normal phenotype. Since enzymes are catalysts and are therefore reusable, one functional allele of an enzyme usually produces sufficient protein to create a wildtype phenotype.
6. Transcription: The nonspontaneous addition of a nucleoside monophosphate to a nucleic acid chain is coupled to the spontaneous hydrolysis of a nucleoside triphosphate into a nucleoside monophosphate and water. Translation: The nonspontaneous addition of an amino acid to a polypeptide and the nonspontaneous translocation of the ribosome are coupled to the spontaneous hydrolysis of GTP.
7. Any of the following: The use of chaperones or chaperonins to facilitate folding. The targeting of misfolded proteins to the proteasome through ubiquitination.

8. Because amyloid fibrils have extremely stable structures and therefore a very low potential energy, so it is very unlikely that they will be able to absorb enough energy to break their intermolecular forces (IMFs) and unfold sufficiently that they could refold properly again.
9. Phenylalanine: hydrophobic, neutral. Serine: hydrophilic, neutral. Leucine: hydrophobic, neutral. Arginine: hydrophilic, basic.
10. a) alpha helices, beta sheets. b) Hydrophobic sidechains on the inside; hydrophilic sidechains on the outside; negatively-charged acidic sidechains next to positively-charged basic sidechains; cysteines form disulfide bonds. c) All of these make ΔH negative by causing the formation of more IMFs/bonds and making IMFs/bonds releases energy: hydrophobic sidechains on the inside maximizes the ability of water molecules and hydrophilic sidechains to form H-bonds; hydrophilic on the outside leads to more H-bonds with water; acidic and basic sidechains form charge-charge IMFs with each other; two cysteines can form a new covalent bond and making bonds releases energy.
11. Enzymes stabilize the transition state by forming IMFs or covalent bonds with atoms in the transition state, thus reducing the potential energy of the transition state.
12. Activate the zymogen to form the active protease structure, likely by cleaving an inhibitory region off of the zymogen. Add acid to create the same pH as the stomach so that the protease can fold into its native shape.
13. a) The bond angles in beta sheets (their 3D shape) makes them more able to aggregate into fibrils than alpha helices. Essentially, sheets can stack against other sheets in stable structures. b) Either of the following: inheriting a mutation in a protein that makes it more likely to fold into a prion form than the wildtype protein; having low levels of chaperones or chaperonins due to advanced age or another factor.
14. d.
15. a, b, c, d, f, h, j.

Unit 3

1. a) Metaphase, b) Metaphase, c) Interphase, d) Anaphase.
2. Circular shape because linear DNA is degraded by exonucleases in bacteria; origin of replication because DNA replication requires it to start; drug resistance gene because if the plasmid isn't actively selected it will be lost over time.
3. Genomic sea urchin DNA contains a higher percentage of A-T base pairs and a lower percentage of C-G base pairs than genomic turtle DNA. Because there are only 2 H-bonds between A and T, but 3 H-bonds between C and G, it takes less energy to break A-T H-bonds than to break C-G H-bonds.
4.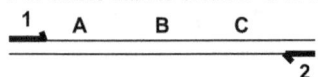
5. a) A ssDNA primer is added to the reaction and the temperature is reduced to below the annealing temperature. b) The temperature is raised the 95°C to denature the DNA.
6. ΔH is negative because H-bonds form when the two ssDNA molecules become dsDNA. ΔS is negative because annealed DNA is more ordered than the separate strands (fewer arrangements are possible when annealed). Therefore, the sign of ΔG depends on temperature. ΔG is negative and annealing happens spontaneously when the temperature is low. ΔG is positive and annealing does not occur (and dsDNA denatures) when the temperature is high.
7. a) PCR and sequencing reactions both contain *Taq* DNA polymerase, dNTPs, a DNA template, and a buffer containing magnesium ions. The concentration of the template must be higher and the complexity of the mixture of DNA template lower for sequencing than for PCR. PCR uses 2 primers, but sequencing uses 1 primer. Sequencing includes a small percentage of labeled, chain-terminating dideoxynucleotides. b) PCR primarily produces products of one length in a typical reaction, and the PCR products accumulate exponentially. Sequencing produces products of every possible length, up to about 1000nt. Sequencing products end with a chain-terminating, labeled dideoxynucleotide and sequencing products accumulate linearly.
8. Any of the following (and probably other answers): Exonucleases could access and degrade the nicked DNA; the chromosomes would break apart when RNA polymerase separated the strands during transcription and genes would be transcribed incompletely; the chromosomes would break apart when helicase separated the strands during the next DNA replication; the damaged DNA would cause the cell cycle to arrest.
9. Proofreading error rate = $(10^{-5})/(10^{-3})=10^{-2}$ or 1% of the time. 100%-1%=99% of the time proofreading happens correctly.
10. 2μL template DNA, 2.5μL primer 1, 2.5 μL primer 2, 25μL master mix, 18μL water.
11. a) Southern blotting requires too large an amount of DNA. Because STR analysis uses PCR to amplify the signal, even a very small amount of DNA can be detected in a forensic sample. b) Southern blotting is preferable because the amount of DNA is not an issue and because Southern blotting is much less likely than PCR to produce false positive results since, unlike PCR, Southern blotting does not have an amplification step.
12. a) The top of the gel should be labeled (-) and the bottom of the gel should be labeled (+). b, i) cannot draw a conclusion – plant DNA could not be detected. b, ii) the food is likely not genetically modified. b, iii) the food is genetically modified.
13. Possibilities include, but are not limited to helicase, RNA primase, ligase, DNA polymerase, RNA polymerase (mostly), most DNA repair enzymes, cohesin, DNA phosphatase.
14. Possibilities include, but are not limited to restriction enzymes, kinetochore proteins, transcription factors.
15. a) The DNA molecule will have the *KanMX* gene in the middle and either side will be sequences homologous to the sites in the genome between which the *KanMX* gene will be inserted through homologous recombination. b) Nonhomologous end-joining; higher mutation rate in *lig4*-deleted because the inability to perform NHEJ would result in less repair of damaged DNA so mutations would be more likely to become permanent.
16. The palindromic sequences that are recognized and cut by many restriction enzymes have the same sequence on both strands. So, it makes sense that each monomer of a dimer would act in an identical way on each of the two strands.
17. Any sequence that is the same when read 5' to 3' on both strands. For example, AAATTT or CAGCTG.
18. a) Add phosphate groups to the 5' ends, either by cutting the ends with a blunt-cutting restriction enzyme or by using phosphorylated primers. b) Treat the plasmid backbone fragment with phosphatase following restriction digestion.
19. They cut viral (bacteriophage) DNA.
20. Bacteria methylate their DNA at restriction sites to prevent cutting.
21. Purify RNA from the white blood cells. Make a cDNA copy of the RNA using reverse transcriptase to create DNA without introns. PCR gene E using primers containing a *Bam*HI site in one primer tail and an *Eco*RI site in the other primer tail. Cut the PCR product with *Bam*HI and *Eco*RI. Cut the plasmid with *Bam*HI and *Eco*RI (optionally phosphatase treat the cut plasmid). Purify the cut PCR product and the plasmid backbone fragment. Ligate these together.
22. a, i) The mutagenesis happened correctly as designed. a, ii) There was template GFP plasmid remaining from the PCR reaction or

there was uncut GFP plasmid contaminating their plasmid backbone preparation. a, iii) A plasmid was assembled, but it was religated plasmid backbone without an insert or there was a mistake in the BFP assembly or a mutation in the PCR of the BFP fragments that eliminated the function of BFP. b) Sequence the BFP gene.

23. (100 million)/(4^8) = 1526 times

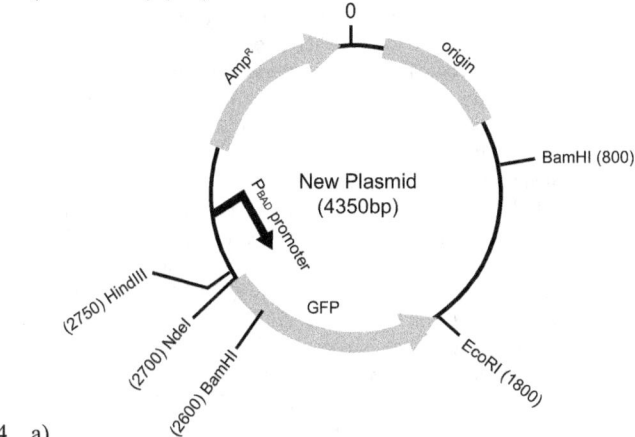

24. a)
    b) Restriction digest or sequencing. c) 1800bp, 2550bp.

Unit 4

1. a, i) eye cells. a, ii) it's the same in both. a, iii) eye cells. a, iv) eye cells. b, i) unmethylated promoter. b, ii) acetylated histones. b, iii) loosely-packed chromatin. b, iv) euchromatin.
2. a) DNA ligase. b) No. The *Bsa*I digests cut the *Bsa*I recognition site off of all of the fragments that are ligated in the Golden Gate reaction. c) Fragments that have complementary sticky ends are ligated to each other, thus determining the order and orientation of the fragments.
3. The bacterial promoter and RBS would need to be replaced with a eukaryotic promoter, 5'UTR and Kozak sequence. The bacterial terminator would need to be replaced with a eukaryotic 3'UTR.
4. The sequence of the promoter determines how well and how often transcription factors and DNA polymerase bind. Methylation at the promoter can prevent the binding of RNA polymerase or transcription factors. Methylation at the promoter can cause the formation of heterochromatin, repressing transcription.
5. No. Drosha is required for the removal of the tails from the microRNA hairpin. This is a necessary step in the production of microRNAs that can be bound by Argonaute.
6. Because the *XIST* RNA is so long, it is still being transcribed, and is physically tethered to the X-chromosome from which it is being transcribed, when it is bound by proteins that also bind to the chromosome. This allows *XIST* RNA to coat only the inactive X chromosome and not the active X chromosome.
7. Because the adult methylation pattern is not fully erased and is not fully restored to the embryonic methylation pattern when the adult somatic nucleus is transferred into the enucleated egg cell. As a result, the gene expression pattern in cloned animals is abnormal.
8. a) The positively-charged sidechains in histone tails attract the negatively-charged phosphate groups in the DNA backbone in both euchromatin and heterochromatin. b) Acetylation of lysine sidechains in histone tails in euchromatin removes some of the positive charges, so DNA packs more loosely because there are fewer positive charges in the histone tails to attract the DNA backbone in euchromatin compared to heterochromatin.
9. It would either encode an RNA that forms a hairpin structure due to base pairing of complementary nucleotides, or it would have two promoters that transcribe the gene in both directions, creating complementary RNAs that could form dsRNA.
10. Following DNA replication, DNA methyltransferase enzymes recognize sites where a CpG dinucleotide is methylated on only one strand and the methyltransferase methylates the CpG dinucleotide on the other strand.
11. The sequence of the 3'UTR affects the stability of the complex that holds together a looped mRNA. Also, microRNAs often bind to the 3'UTRs of mRNAs to trigger their uncapping and deadenylation.
12. a) The gene is imprinted in that the maternal allele is silenced. A person has the disease phenotype if they inherit a disease allele from their father and their mother's wildtype allele is silenced. A person does not have the disease phenotype if they inherit a wildtype allele from their father, regardless of the type of allele they inherit from their mother. b) zero. c) 25%.

Unit 5

1. Because *p53* functions in multiple pathways that regulate the cell cycle and apoptosis, so mutating *p53* tends to be more advantageous to the tumor than mutations in genes that affect only one pathway. The mutation of *p53* tends to do more to make the tumor malignant than the mutation of most other genes.
2. Telomerase is expressed in the germline because it is important that telomeres do not shorten from one generation of person to the next. However, the lack of telomerase expression in somatic cells can limit the number of times that somatic cells can divide, which is thought to be a tumor-suppressive mechanism.
3. Childhood cancers arise in fetal cells in the fetus or in early childhood and fewer genes tend to be mutated than in adult cancers. Adult cancers require mutations to occur in a larger number of genes and they arise in adult cells, usually later in life.
4. a) VDJ recombination assembles one V, one D, and one J segment from many possible segments, creating antibody diversity that is enhanced by introduction of additional mutations at the segment junctions. There is negative selection against antibodies that recognize self. b) Somatic hypermutation generates additional antibody diversity after an antibody has recognized an antigen. There is positive selection for antibodies that recognize the antigen and that positive selection is stronger for antibodies that bind the antigen better.
5. a) #1 and #3. b) Because they share the fewest number of blocks of conserved synteny, this means that the individual blocks are larger and thus there has been the least shuffling between these two genomes.
6. The copy of a gene created by gene duplication is extra and is not needed to create the normal phenotype. Therefore, one of the gene copies is not subject to selection and mutations that it acquires are not selected against as long as they don't create a new, harmful phenotype. Over time, the gene copy might accumulate enough changes that it gains a novel beneficial function that is positively selected.
7. Telomeres have special TG-rich DNA sequences that are bound by specific telomere end-binding proteins. These telomere proteins distinguish a telomere from a DNA break.
8. a) T. b) O. c) O. d) T. e) O. f) T. g) T. h) O.
9. The heat triggered the unfolding of the Cdc13-1 protein, causing it to come off of the telomere. Exonucleases treated the telomere like a DNA break and degraded the chromosome. The WT Cdc13p and the Cdc13-1 protein that had not been heat-shocked were properly folded, bound to the telomere, and prevented DNA repair proteins like exonucleases from binding at the telomere. B) The loss of the *ADE5* gene in *can1* mutants could be observed when the yeast changed from red to white and this color change in canavanine-resistant colonies indicated that both *ADE5* and

CAN1 had been lost, suggesting a large deletion rather than a point mutation.
10. a) Dominant because too many unrelated people (8) would need to carry a disease allele in order to explain recessive inheritance, but only one unrelated person needs to have the disease allele to explain dominant inheritance. b) Since it is a childhood cancer, not every cell in the child's body has the cancer phenotype, suggesting that a mutation in the wildtype allele must be acquired to actually create a tumor. It's likely that any person with the disease phenotype in the pedigree inherited one mutant allele in all of their cells and then happened to acquire a mutation in the other allele by chance at least one place in their body. c) Tumor suppressor gene. d) Person #18 inherited a mutant allele, but happened to not develop a mutation in the wildtype allele in any cells. This person passed the mutant allele down to his affected child.

Unit 6

1. A long-read technique is preferable when sequencing a genome for the first time because longer sequences can be assembled more reliably than short sequences, which is important when a reference genome is not available. So far, short read techniques can be done more rapidly and more cost effectively, which makes them preferable when resequencing a genome where sequences can be compared to a reference sequence.
2. Sanger sequencing is the only technique where only one sequence can be obtained for very little cost.
3. Genome editing could be directed against the HIV sequence since HIV integrates into the genome. Cutting of the HIV sequence and imprecise repair could abolish HIV's activity. Another strategy is to use genome editing to target a patient's *CCR5* gene. Since the cell surface CCR5 protein is required for HIV entry into cells, its deletion could prevent HIV from infecting new cells. There is no known adverse effect to going without CCR5 protein. This strategy could have an advantage over the strategy to target HIV since the HIV could still spread through infection of new cells if some copies escaped the genome editing.
4. Early adenoviruses overactivated the immune system and could cause shock that killed the patient. This was overcome by replacing the viral surface proteins that activated the immune system with versions of those proteins that did not activate the immune system. Also, modified adeno-associated viruses that are less immunogenic have also been used. Early gamma retroviruses caused leukemia at a high rate by integrating near oncogene promoters and triggering the transcription of those oncogenes. Newer viruses don't integrate near promoters and the viral sequences that activated transcription have been eliminated.
5. TALENs are easier because the code whereby amino acids recognize specific DNA nucleotides is much more regular and predictable for TALENs than for zinc finger endonucleases. CRISPR/Cas9 is even easier than TALENs because the specificity of CRISPR/Cas9 is conferred by the sequence of the CRISPR RNA, which is easy for the researcher to design (since complementarity between DNA and RNA is clear) and build (since oligonucleotides can be easily ordered from DNA synthesis companies). The engineering of a TALEN protein is much more involved.
6. In a recessive disease or in a dominant haploinsufficient disease, you would want to put the wildtype gene back in the genome. You could either use gene therapy to introduce a new copy of the gene or you could use CRISPR/Cas9 to cut the disease allele and use a DNA patch and homologous recombination to replace the disease allele with wildtype sequence. In a gain of function or dominant negative disease, you would want to delete the mutant allele and it might not matter whether you replaced it with the wildtype allele. You could probably rely on NHEJ to imprecisely "repair" a CRISPR/Cas9 cut.

Appendix 2: Basic Molecular Biology Techniques

Using a Centrifuge

A centrifuge spins samples at high speed, creating a force on the sample that is greater than the force of gravity. This is used to separate components of a mixture based on density.

Protocols will usually refer to the g-force at which you should spin your sample. Spinning at 10,000 x g (10,000 times the force of gravity) or higher is typical for many applications. Your centrifuge might allow you to set it based on the g-force (rcf = relative centrifugal force), or it might require you to set it based on rpm (revolutions per minute). The relationship between rpm and rcf depends on the radius (r) of the rotor in centimeters, so it will be different for each centrifuge: $rcf = (0.00001118)(r)(rpm)^2$

When putting tubes in the centrifuge they must always be balanced. Running an unbalanced centrifuge can cause the rotor to shake and damage the centrifuge. In extreme cases, an unbalanced rotor can even fly off, which is particularly dangerous when using large, high speed centrifuges.

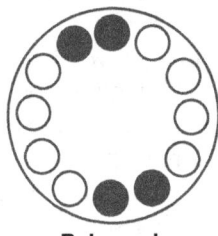
Balanced

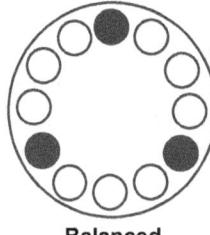
Balanced

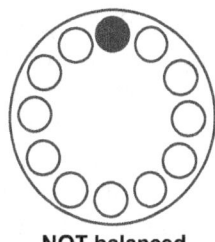
NOT balanced

Gel Electrophoresis of DNA

Gel electrophoresis is the separation of biological molecules based on size and charge. Negatively charged DNA starts in the wells near the negatively charged (black) electrode and migrates toward the positively charged (red) electrode. Small fragments migrate more quickly than larger fragments. DNA fragments "run to red." DNA samples are mixed with loading dye at a final concentration of 1X prior to loading. A DNA ladder is also loaded in at least one well. The DNA is visualized because the gel is stained with a dye that binds DNA. The dangerous mutagen ethidium bromide is the most commonly-used dye because it is relatively inexpensive and highly sensitive and has been around the longest. DNA bound by ethidium bromide fluoresces under UV light. Safer green dyes are less mutagenic and allow DNA to be visualized under blue light. However, gloves should always be worn when handling anything containing a DNA stain because of the potential for any molecule that binds DNA to act as a mutagen. A great animation explaining gel electrophoresis is available here: http://learn.genetics.utah.edu/content/labs/gel/

A standard DNA gel is 1% agarose in 1X running buffer. A higher percentage of agarose will resolve smaller fragments better. A lower percentage agarose will resolve larger fragments better. You can find tables online of optimal percentages of agarose depending on the sizes of the fragments you want to resolve and the type and concentration of running buffer. There are different types of running buffer (TBE, TAE, TTE, sodium boric acid) that differ in terms of how they resolve DNA, their stability, and the maximum current that they can withstand. I tend to use TAE if I need very good, publication-quality resolution, but I routinely have students use sodium boric acid (Brody and Kern, *Biotechniques* 2004, https://doi.org/10.2144/04362BM02) in teaching labs because it can tolerate a very high voltage which facilitates students having the time to load, run, and photograph gels within a single class period.

Notes on Mixing

In the reactions described below, buffers are used that contain appropriate salts and sometimes other compounds at an appropriate pH. Because these buffers are stored in the freezer, some of them have a tendency to precipitate out of solution. It's important to mix any buffer well before use and to keep mixing as long as any precipitate is visible. If the buffer doesn't contain any enzymes or other proteins, it can and should be vortexed.

Any stock vial or reaction mixture containing an enzyme should never be vortexed because vortexing can physically denature enzymes. An enzyme stock vial should not need mixing, but it may need to be centrifuged to make sure all of the enzyme is collected at the bottom of the tube. It's generally wise to add enzymes last to reaction mixtures so that the reaction already contains buffer at the right concentration so that the enzyme continues to fold properly in the reaction mix. When a reaction mixture is complete, the microcentrifuge tube can be flicked with your finger to mix the contents and then spun down in a centrifuge. Some labs lack the equipment to easily spin down 0.2mL PCR tubes, so in that case, just be sure that all of the reaction components get into the bottom of the tube as you add them.

Pay attention to how protocols ask you to mix solutions and follow the suggestions. Vortexing is significantly more powerful, but also harsher than flicking a tube, pipetting up and down, or inverting (turning tubes upside down). Pipetting up and down can be especially helpful when trying to resuspend a pellet of bacteria or DNA into a solution because the pipet tip can help to dislodge and stir a dense pellet.

It's always important to watch the liquid in your pipet tips to make sure that you're dispensing what you intend to deliver. Even when you're an expert this is important because it's easy to unintentionally pipet air. Especially with volumes less than one microliter, you should look closely to make sure you see liquid in the pipet tip before dispensing and you should put the pipet tip under the surface of the liquid in the tube you're dispensing into when expelling a very small volume of liquid to be sure that you actually transfer it. When combining multiple solutions, it's best to add any solution with a very small volume only after there's a substantial volume of other solution already in the new tube. If you are working with a partner in a class laboratory, it's much better for the person doing the pipetting to actually hold the tube (than for one person to hold the tube while the other person pipets) so that you can pay attention to all of the above (and because it's also easier to get the tip into the tube when you have control of both items).

PCR

PRIMER DESIGN

1. There are various computer programs to facilitate primer design and primer analysis, but the directions below are how to do it yourself – and the principles are important even if you use primer design software.
2. The two primers should ideally be 20-22 bp. 18bp is probably the absolute minimum, but should be avoided. Longer than this is better than shorter because you lose specificity if the primers become too short.
3. If you have some choice of where to locate your primer, choose an area of the DNA that is nonrepetitive in sequence and that ideally has a mixture of all four nucleotides.
4. The annealing temperatures (T_m) of the two primers must be within a couple of degrees of each other.
5. Annealing temperatures should be between 55°C and 65°C. I've had good luck designing primers that are at 60°C using the [$T_m = (4°C)(\text{\# of Cs or Gs}) + (2°C)(\text{\# of As or Ts})$] method and then using an annealing temperature of about 55° in the thermal cycler.
6. If designing primers that have extra 5' tail sequence (such as if adding a restriction site) do not count these extra nucleotides when calculating the T_m. If adding on a tail with a restriction site, add an extra 8nt to the 5' end (5' to the restriction site) because many restriction enzymes can't cut at the extreme 5' end of a sequence. You can consult the NEB catalog for the precise number of extra nucleotides needed for each enzyme.
7. Try to make the last one or two of the 3' nucleotides G or C for better annealing where DNA polymerase starts. (I'm not completely sure this actually helps, but I was taught to do it and have always had good luck with my primer design – a case of my never having changed or tested it because it works…a phenomenon that happens often in labs).
8. Check for the ability of the primers to form hairpin structures or primer dimers (3' ends shouldn't be able to anneal to each other).
9. If PCRing from a complex genomic template, BLAST your primers to check that they won't amplify other sites.

PCR REACTION SETUP

General PCR using a Master Mix containing *Taq* polymerase, dNTPs and buffer:

Component	Stock Concentration	Final Concentration	Sample volumes for a 50μL rxn
Forward Primer	5μM	0.2 μM	2 μL
Reverse Primer	5μM	0.2 μM	2 μL
Template DNA	varies	0.02ng/μL-20ng/μL genomic DNA* or 0.02pg/μL-20pg/μL plasmid DNA	(5 μL) varies
2X Master Mix	2X	1X	25 μL
Water for PCR	n/a	add up to final volume (25-100 μl)	(16 μL) varies

*2 ng/μL yeast genomic DNA works well. Use less for an organism with a smaller genome and more for an organism with a larger genome.

Colony PCR using a Master Mix:

Component	Stock Concentration	Final Concentration	Sample volumes for a 100μL rxn
10 μM Forward Primer	5μM	0.2 μM	4 μL
10 μM Reverse Primer	5μM	0.2 μM	4 μL
2X Master Mix	2X	1X	50 μL
Water for PCR	n/a	add up to final volume (50-100 μl); more is better	42 μL
Bacteria	n/a	Touch a pipet tip to a bacterial colony then dip into PCR mix; swirl (less bacteria is better)	n/a

Controls for PCR

Always include a negative control reaction tube lacking DNA template. This is critical to check for contamination of your reactants.

Sometimes you may use PCR to check for the presence or absence of a particular DNA sequence. A good strategy is to include a positive control reaction in the same reaction tube as your diagnostic PCR. Design two other primers that should definitely amplify a product in your template DNA. Therefore, you should expect to see the amplification of both the control and diagnostic bands or only the amplification of the control band. PCR tends to amplify shorter products more efficiently than larger products, so making your control band larger than your diagnostic band will add confidence that the template for the diagnostic band really wasn't there if you don't see it but do see the larger control band.

Appendix 2: Basic Molecular Biology Techniques

CYCLING CONDITIONS FOR PCR

Always use a hot start (heat the PCR machine to 95°C and transfer tubes directly to the heated block from ice). This prevents the PCR reaction from starting during the setup, which can create unwanted side products due to off-target annealing of primers. Any unwanted side product will be amplified in the PCR reaction because it contains the primer binding sites.

General PCR

Cycles	Temperature	Time	Notes
	95°C	5 min	Denatures template completely
Repeat 35 times:	95°C 45-65°C 72°C*	30 sec 30 sec 1 minute per kb of template	Denaturation Primer annealing Elongation
	72°C*	5 min	Extra elongation for completing products
	4°C	∞	Prevents enzyme activity

* 72°C is optimal for most *Taq* polymerases, but consult the product literature because some are different.

PCR for primers with tails

Cycling	Temperature	Time	Notes
	95°C	5 min	
Repeat 15 times:	95°C temp1: (45-65°C) 72°C*	30 sec 30 sec 1 minute per kb of template	Uses the annealing temperature for the primer excluding the tail. Used to start to create product.
Repeat 35 times:	95°C temp1 + 5°C 72°C*	30 sec 30 sec 1 minute per kb of template	Uses a higher annealing temperature because products containing the full primer sequences are the templates. A higher annealing temperature increases specificity of annealing.
	72°C*	5 min	
	4°C	∞	

Colony PCR of bacteria

For the first step, use 10 min at 95°C instead of 5 min at 95°C. This allows better lysis of bacteria and denaturation of bacterial enzymes. The other cycling conditions are normal.

Restriction Enzyme Digests

After deciding which enzyme(s) to use in the digest, look them up in the company's catalog (I have always used NEB) to determine the reaction temperature for each enzyme, the ideal buffer to use with each enzyme, and whether or not one of the enzymes requires BSA in the reaction mix. If your enzyme is methylation-sensitive, double check that your desired cut sites are not methylated. If the cut sites are methylated, DNA must be extracted from methylase-deficient bacteria.

Decide how much DNA to digest. Digesting 5μL of miniprep DNA is sufficient for screening colonies. When digesting DNA to use a fragment in a ligation reaction, be sure that you are digesting enough DNA that the fragment you need will be concentrated enough after gel extraction to be used in your ligation reaction.

A Restriction Digest with One Enzyme

Component	Stock Concentration	Final Concentration	Sample volumes for a 20µL* rxn
NEBuffer	10X	1X	2 µL
BSA	100X	1X or none	(0.2 µL) or 0µL
DNA	varies	Varies (5µl miniprep DNA to a few µg of DNA)	(5 µL) varies
Restriction Enzyme	5000-20000 Units/mL	5-10 Units	0.5 µL (cannot exceed 10% of the total rxn volume)
Water	n/a	add up to final volume (20-50µL)	(12.3 µL) varies

A Restriction Digest with Two Enzymes

Component	Stock Concentration	Final Concentration	Sample volumes for a 20µL rxn
NEBuffer	10X	1X	2 µL
BSA	100X	1X or none	(0.2 µL) or 0µL
DNA	varies	Varies (5µl miniprep DNA to a few µg of DNA)	(5 µL) varies
Restriction Enzyme 1	5000-20000 Units/mL	5-10 Units	0.5 µL (cannot exceed 5% of the total rxn volume)
Restriction Enzyme 2	5000-20000 Units/mL	5-10 Units	0.5 µL (cannot exceed 5% of the total rxn volume)
Water	n/a	add up to final volume (20-50µL)	(11.8 µL) varies

*A 20µL reaction volume is sufficient for analyzing miniprep DNA. If digesting larger quantities of DNA for cloning purposes, use a 50µL reaction volume.

Flick the tube to mix and spin down in a microcentrifuge. Incubate the reaction for 2 hours at the optimal reaction temperature. If digesting plasmids to create fragments for cloning, it's important that the reaction goes to completion. Spinning down the reaction at some point during the incubation is helpful to be sure all of the plasmid has been exposed to the enzymes in the reaction mix and to recollect any condensed liquid under the cap back in the bottom of the tube to make sure the concentrations of the reaction components are optimal. If possible, putting tubes in an incubator rather than in a heat block helps to minimize condensation under the cap.

If reactions with two enzymes require different temperatures, the reactions will need to be performed sequentially. If the two enzymes have different ideal buffers, the restriction enzyme catalog should give you information about whether there is a buffer that will work for both enzymes. If not, the reactions will need to be performed sequentially. If only one of the two enzymes requires BSA, include BSA in the reaction mix.

Phosphatase Treatment of DNA

CIP, calf intestinal phosphatase, will remove 5' phosphate groups from DNA ends. This should be done to a vector backbone fragment used in cloning if it has complementary sticky ends, blunt ends, or if colonies are produced in a "vector only" ligation. CIP is active in many restriction enzyme buffers and can usually be added when a restriction digest is set up. Consult the company literature for protocols.

Gel Purification of DNA

1. After digesting (and phosphatase-treating) DNA, add loading dye to samples (1X final concentration) and run the samples on an agarose gel.
2. Photograph the gel.
3. Place the gel on a glass plate. If using ethidium bromide, illuminate the gel with UV light, wear UV-blocking goggles to protect your eyes, and use a razor blade to cut out the desired DNA bands. UV light can create breaks and other DNA damage in the DNA so minimize the time that the UV light is on and use a long-wave UV light rather than a short-wave light if possible. If using a green DNA stain, illuminate the gel with blue light and wear orange glasses when cutting out bands. The visible blue light is not a danger to you and will not damage the DNA. Use a new razor blade to cut out each desired gel slice. Avoid touching the razor blade to undesired DNA bands to avoid cross-contamination. Trim away as much extra agarose as possible so that the slice contains just the DNA. Usually 5 cuts are necessary – a vertical cut right next to the DNA on each of the 4 sides of the band, then turn the slice on its side and trim off the agarose on the bottom.
4. There are several commercially-available kits for purifying DNA from gel slices. They tend to involve adding a buffer and heating the mixture so that the agarose dissolves in the buffer. The dissolved solution is run through a spin column that collects the DNA which is then washed with an ethanol solution and eluted from the spin column. When using these kits, it is critical to follow the directions to remove all of the ethanol because ethanol will inhibit subsequent reactions.

Ligation of DNA

There are several commercially-available DNA ligases, each of which has its own protocol. When setting up ligation reactions, you should always set up an extra negative control reaction that contains the same concentration of the vector backbone fragment as your experimental ligation, but that does not include the insert. This control will allow you to estimate how much unwanted background you have in your ligation reaction. If you get the same number of colonies from your "vector only" ligation as from your experimental ligation, you will know that the background of incorrect ligations is too high to allow you to find a correct clone. In this case you'll have to repeat some or all of the previous steps.

Transformation

There are two general types of transformations, low efficiency and high efficiency. A low-efficient transformation is sufficient for intact circular plasmids, which can be much more efficiently transformed than ligation reactions. These protocols allow bacteria from a fresh colony to be soaked in calcium chloride along with DNA for 30 minutes and then heat shocked, grown out and plated. A high efficiency transformation is necessary for ligation reactions. Chemically competent cells are either purchased from a company or can be prepped in a lab if you have a refrigerated centrifuge, a 4°C cold room, and are willing to work for long periods of time in the cold room. In both the low and high efficiency transformations, after incubating the competent cells with the DNA for 30 min, the bacteria are heat shocked at 42°C for about 30-60 seconds, depending on the size and shape of the tube. The bacteria are transferred back to ice for a few minutes. Bacteria can be plated then for the low efficiency transformation or grown in liquid growth media for 30 min prior to plating for the high efficiency transformation.

Miniprep of plasmid DNA

A miniprep is the preparation of plasmid DNA from a small volume of bacterial liquid culture (2-5 mL). There are several commercially-available miniprep kits and also protocols for doing minipreps without a kit. In my experience, the quality of the DNA you can get from a kit is much higher (and the protocol is faster and easier) than if minipreps are done without a kit. These protocols all start with pelleting the bacteria and then resuspending the pellet in a fairly neutral solution that is isotonic to the bacteria and has optimal pH and salts. A solution containing NaOH, a strong base, and detergent is added next to lyse the bacteria. Following a brief incubation, a buffer solution containing potassium acetate and acetic acid that neutralizes the lysis solution is added. This step also precipitates the proteins and lipids, leaving the plasmid DNA in solution. The bacterial genomic DNA (large circular chromosome) is attached to the cell membrane and precipitates with the lipids. The mixture is then centrifuged for several minutes to pellet the precipitate. The supernatant containing the plasmid DNA is processed further using techniques that purify the plasmid DNA away from the rest of the contents of the supernatant. If using a commercial kit, the DNA is collected and washed in a spin column. If not using a kit, the plasmid DNA is precipitated by adding ethanol at a final concentration of 70% or by adding isopropanol at a final concentration of 50%. DNA is insoluble in alcohol. The precipitated DNA is pelleted by spinning for up to 30 min in a cold microcentrifuge and the supernatant is removed. Protocols often include a second wash with alcohol solution and a second centrifugation to collect the precipitated DNA. After removing the supernatant, the pellet is air dried to remove residual alcohol (critical because alcohol inhibits enzymes). Finally, the purified DNA is resuspended in water. At some point in the protocol, RNase A is added to degrade RNA. When pelleting DNA, be sure to position the hinge of the tube outward in the centrifuge. DNA pellets are often very small and hard to see, but if you do this, you'll know that the pellet is close to the bottom of the tube directly under the hinge. When removing supernatant, avoid touching the pellet. If you can see the pellet, be careful not to dislodge it. DNA pellets are easier to see when they are less pure and contain a lot of salt – these pellets are white and fluffy. When DNA pellets are more pure, they are clear and glassy.

Appendix 3: *E.coli* and Phage Promotors

A promoter is a DNA sequence where RNA polymerase recognizes specific sequences as it binds to the DNA to initiate transcription of a downstream RNA transcript. A promoter often also contains binding sites for additional transcription factor proteins that positively or negatively affect the binding or activity of RNA polymerase.

In bacteria, RNA polymerase is a complex of proteins, including the core enzyme that catalyzes the transcription reaction and a protein called sigma factor that helps to recognize specific "-10" and "-35" elements in the promoter DNA, as described in the figure below.

E. coli promoter elements
Consensus sequences and spacing of -35 and -10 elements are shown. Transcription begins at +1.

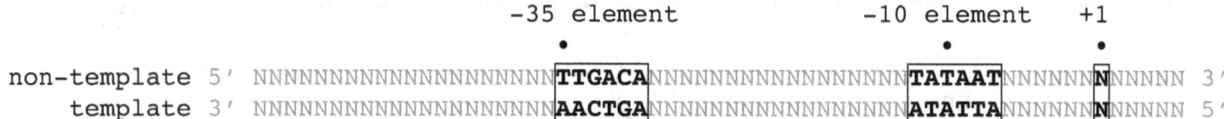

Sigma factor and the core RNA polymerase initiating transcription at the promoter
Sigma factor (light gray) binds to the -35 and -10 elements and unzips the DNA beginning at the -10 element. Unzipping continues past the +1 transcription initiation site. The template strand is positioned within the active site of the core enzyme (medium gray) which includes an Mg^{2+} ion.

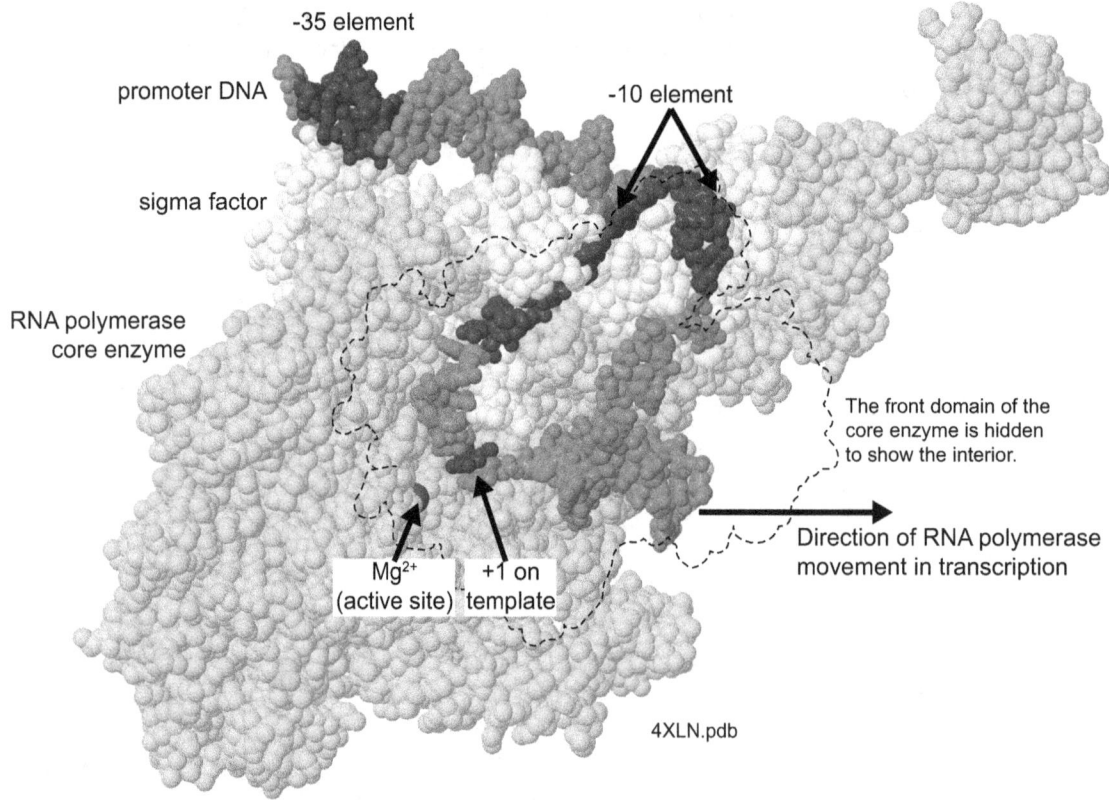

Several examples of inducible and repressible promoters are described below for reference when you encounter these promoters in genetic engineering. All of these are transcribed by *E.coli* RNA polymerase except for the T7 bacteriophage promoter. The *trp* and *lac* promoters are classic textbook examples of repressible and inducible promoters, respectively. The promoters below are outlined in their normal context in the *E. coli* or phage genome and their manipulation in genetic engineering is discussed. EcoCyc.org provides updated summaries of the functions of *E. coli* genes and their promoters and also provides multiple references: Keseler, I. M., *et al.* EcoCyc: fusing model organism databases with systems biology, *Nucleic Acids Research* 41, D605–D612 (2013).

Trp Promoter

In *E. coli*, the *trp* promoter controls the expression of the *trpEDCBA* operon containing five ORFs encoding enzymes that work together to synthesize the amino acid tryptophan (EcoCyc.org TU00067). A second gene, *trpR*, encodes the Trp repressor, TrpR protein, which forms dimers, but does not bind DNA when tryptophan is absent. The *trp* operon is transcribed when tryptophan is absent. The binding of tryptophan to the TrpR dimer helps it fold into a more stable shape that can bind DNA. TrpR binds as three adjacent dimers to the *trp* promoter when tryptophan is present, preventing the binding of RNA polymerase and thus preventing transcription.

Transcription from the *trp* promoter in the absence of tryptophan

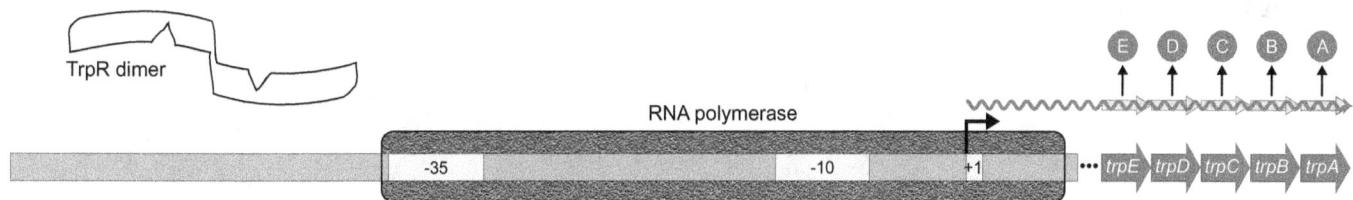

Transcription from the *trp* promoter is repressed by tryptophan and TrpR

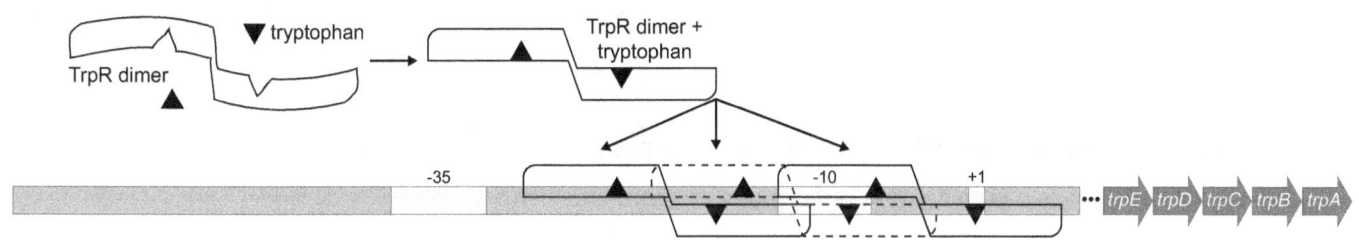

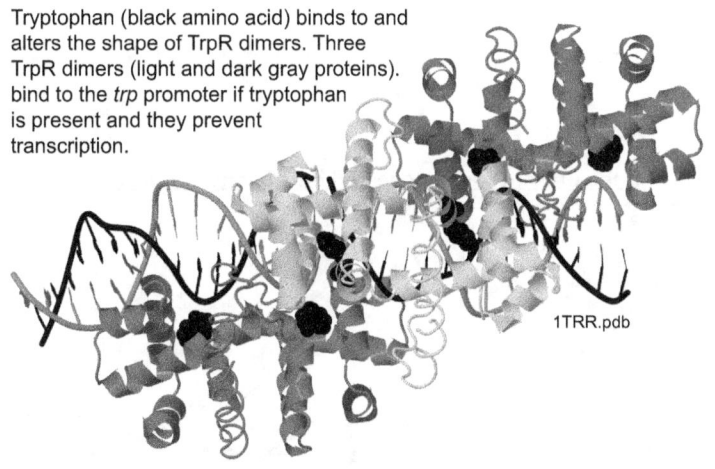

Tryptophan (black amino acid) binds to and alters the shape of TrpR dimers. Three TrpR dimers (light and dark gray proteins) bind to the *trp* promoter if tryptophan is present and they prevent transcription.

1TRR.pdb

Lac promoter

In *E. coli*, the *lac* promoter controls the expression of the *lacZYA* operon containing three ORFs encoding proteins that work together to take in and catabolize the sugar lactose (EcoCyc.org TU00036). A second gene, *lacI*, encodes the *lac* repressor protein, which binds to the *lac* promoter and represses transcription when lactose is absent. The *lac* repressor forms a dimer and two dimers bind two separate binding sites in the *lac* promoter, one of which overlaps with the binding site of RNA polymerase. This creates a loop in the promoter DNA that facilitates the repression of transcription by constraining the conformation of the intervening DNA in a shape that prevents the binding of RNA polymerase to the promoter. When lactose is present, some of it is converted into allolactose, which has a slightly different structure. The binding of allolactose to the *lac* repressor makes it fold into a shape that no longer binds DNA. Additionally, the *lac* promoter has binding sites for the catabolite gene activator protein (CAP), which activates transcription at many promoters by helping to activate RNA polymerase through interactions with one of RNA polymerase's C terminal domains. CAP binds to the *lac* promoter only when cyclic adenosine monophosphate (cAMP) levels are high due to glucose levels being low. CAP is also called the cAMP receptor protein (CRP). CAP binding at the promoter not only enhances transcription when lactose is present, but it also enhances the repression of transcription by the *lac* repressor when lactose is absent by further stabilizing the loop structure. Therefore, transcription of the *lac* operon is highest when lactose is present and glucose is absent.

Transcription from the *lac* promoter is repressed in the absence of lactose by the Lac repressor protein

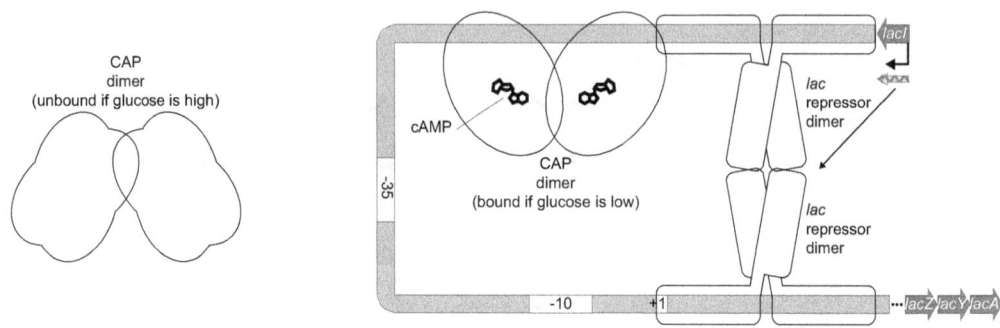

Transcription from the *lac* promoter is not induced in the presence of lactose and the presence of glucose

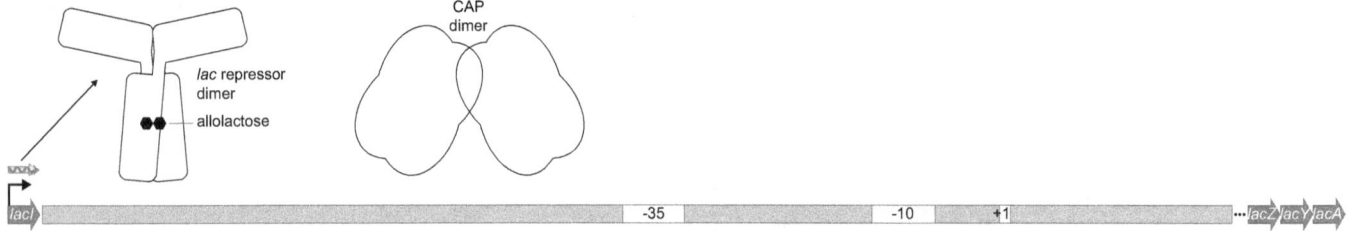

Transcription from the *lac* promoter in the presence of lactose and the absence of glucose

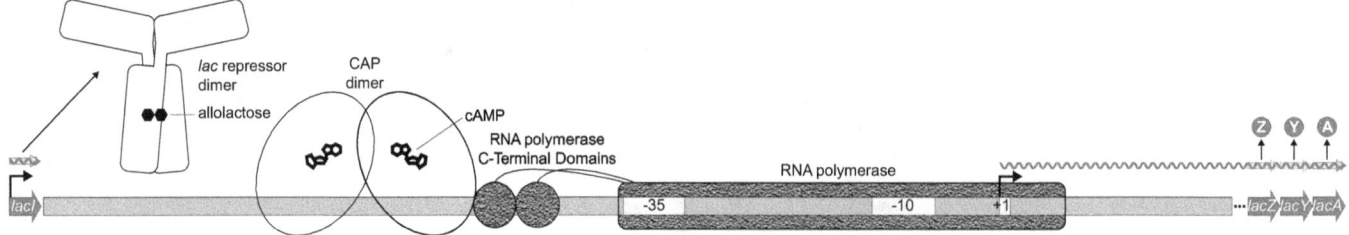

In engineered systems including the *lac* promoter, IPTG, a molecular analog of allolactose, is used to induce expression from the *lac* promoter by inhibiting the *lac* repressor. IPTG is used instead of lactose because unlike lactose, IPTG cannot be catabolized by the product of the *lacZ* gene, beta-galactosidase. The *lac* promoter can only be used in cells that express the *lac* repressor. Additionally, the *lac* promoter used in engineered systems (*lac UV5*) contains point mutations in the -10 element and elsewhere that substantially decrease its dependence on CAP binding, so that glucose and cAMP levels have little impact on *lac* expression. Therefore, the addition of IPTG alone is sufficient to induce transcription.

Tac Promoter

The engineered *tac* promoter is an artificial fusion of the *trp* and *lac* promoters that is repressed by the *lac* repressor protein, but is expressed about 10-fold more than the *lac* promoter when induced by IPTG. The version of the *lac* promoter used in this fusion is *lac UV5*, which is not dependent on CAP binding for expression. This promoter must be used in cells that express the *lac* repressor if it will be used for IPTG induction. If this promoter is used in cells that do not express the *lac* repressor, it will be on constitutively.

Transcription from the *tac* promoter is repressed in the absence of lactose by the Lac repressor protein

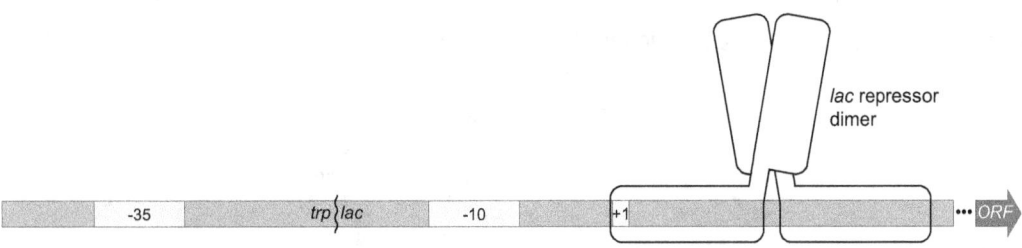

Transcription from the *tac* promoter in the presence of lactose

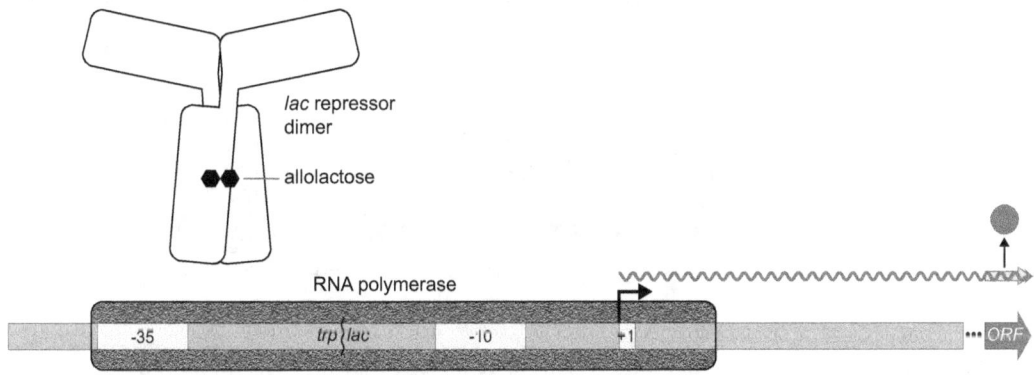

De Boer, H., Comstock J. L. & Vasser, M. The *tac* promoter: a functional hybrid derived from the *trp* and *lac* promoters. *PNAS* 80, 21–25 (1983).

P_{BAD} promoter and AraC

The *araBAD* operon (EcoCyc.org TU00214) contains ORFs that encode proteins that catabolize the sugar arabinose. The *araC* gene encodes the AraC protein which acts as both a repressor and inducer of transcription. The *araC* promoter, Pc, and the *araBAD* promoter, P$_{BAD}$ are overlapping and initiate transcription in opposite directions. When there is no AraC protein, transcription of *araC* occurs from the Pc promoter, but there is no transcription from the P$_{BAD}$ promoter. When AraC protein is present, but arabinose is absent, an AraC dimer binds to the promoter region in an "open" conformation in which it binds to one binding site in the Pc promoter and one binding site in the P$_{BAD}$ promoter, forming a loop in the DNA that represses transcription from both the Pc and P$_{BAD}$ promoters. In the presence of arabinose, arabinose binds to AraC and AraC dimers adopt a "closed" conformation. One dimer binds to adjacent binding sites in the Pc promoter, repressing transcription from the Pc promoter by overlapping the binding site for RNA polymerase. A second AraC dimer binds to adjacent binding sites in the P$_{BAD}$ promoter, which promotes the binding and activation of RNA polymerase at the P$_{BAD}$ promoter. Therefore, AraC protein represses transcription from the P$_{BAD}$ promoter in the absence of arabinose, but induces transcription from the P$_{BAD}$ promoter in the presence of arabinose. As discussed above for the *lac* promoter, CAP-cAMP binding to the P$_{BAD}$ promoter also helps to induce transcription from the P$_{BAD}$ promoter, so glucose can repress transcription from the P$_{BAD}$ promoter.

Transcription from the Pc promoter in the absence of AraC protein

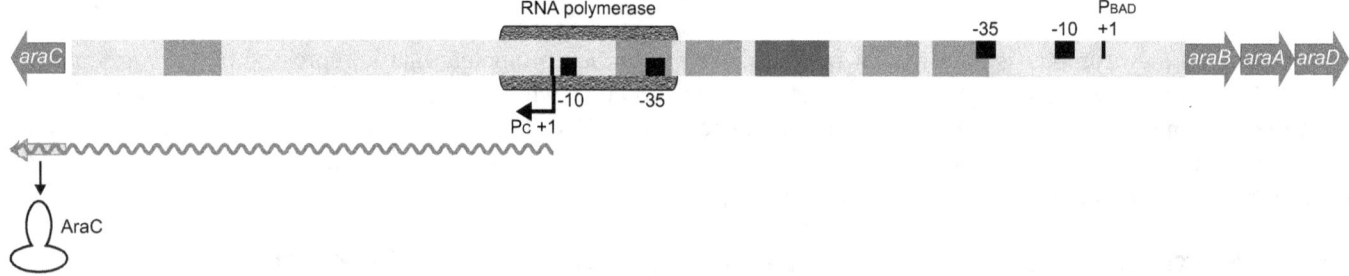

AraC represses transcription from the Pc and the P_{BAD} promoters in the absence of arabinose

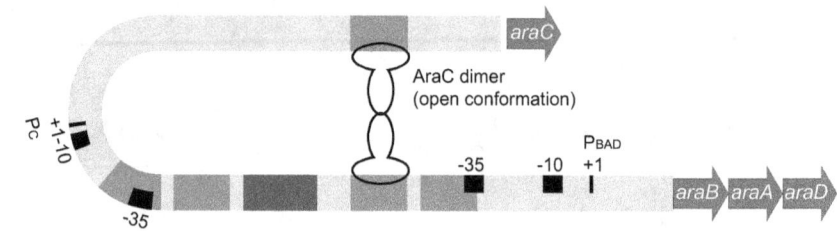

AraC represses transcription from the Pc promoter and induces expression from the P_{BAD} promoter in the presence of arabinose

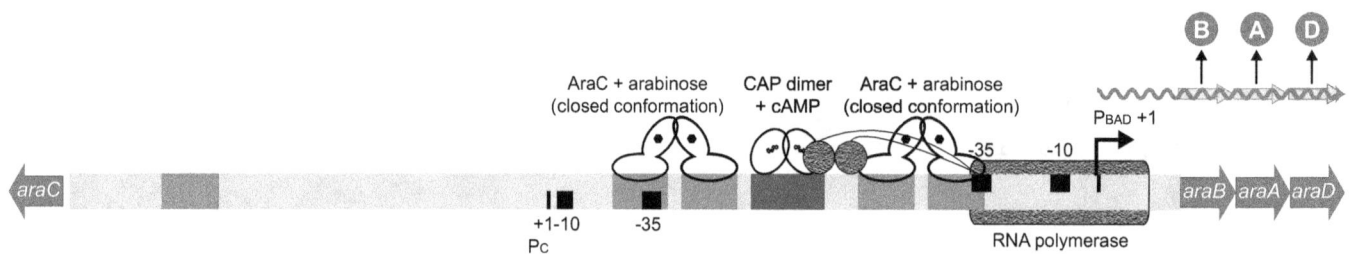

When used for engineering, "P_{BAD}" plasmids contain the P_{BAD} promoter as well as the full *araC* gene. P_{BAD} plasmids only work in some genetic backgrounds. The bacterial strain must be capable of transporting arabinose into the cell (it must have genes *araEFGH*), but not capable of catabolizing arabinose (it must lack genes *araBAD*, and generally also lacks *araC* which is supplied on the plasmid). For example, TOP10 and DH10β bacteria fit these criteria, but DH5α bacteria do not.

Lambda P_L and P_R promoters

Bacteriophage lambda infects *E. coli* and can either have a lytic phase of its life cycle where it expresses its genes and replicates to produce more viruses that lyse the host bacterium, or it can have a lysogenic phase of its life cycle where it integrates into the *E. coli* genome and does not reproduce. The ability to maintain a lysogenic state depends upon the repression of transcription from two lambda promoters, P_L and P_R, that direct the expression of multiple lytic viral genes on the left and right sides of the viral genome, respectively. Transcription from the P_L and P_R promoters can be repressed by binding of lambda repressor protein dimers to three sites within each of those promoters, two of which overlap with the -10 and -35 elements. This binding blocks access of RNA polymerase to the promoter. The lambda repressor protein is encoded by the *cI* gene, which is the only gene that is expressed from the lambda genome during the lysogenic phase. The regulation of the switches between the lytic and lysogenic phases of the lambda life cycle is a classic textbook example of genetic regulation and there are significant details that are not discussed here.

Transcription from the lambda P_L and P_R promoters is repressed in the presence of lambda repressor

Promoters P_L and P_R are located near each other, but at different loci in the phage lambda genome, and they initiate transcription in opposite directions, to the left and right, respectively. However, the mechanism of their regulation is very similar. The -35 element, -10 element, and transcription start site are shown in bold and underlined for each promoter. The binding of dimers of lambda repressor protein to the lambda repressor binding sites (dashed rectangles) represses transcription both promoters.

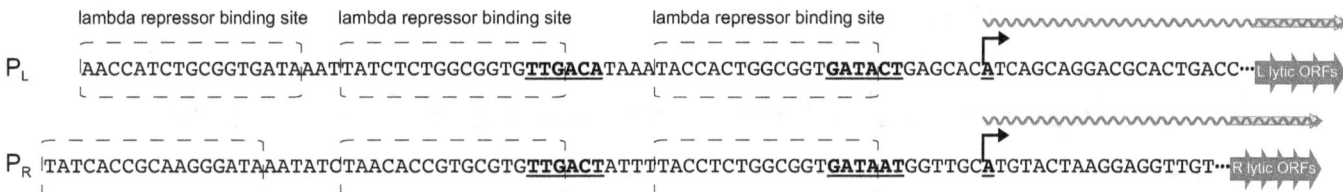

Both the P_L and P_R promoters have been used in genetic engineering, often including only two of the three lambda repressor binding sites in an engineered promoter. For engineering purposes, a temperature-sensitive mutant of the lambda repressor, *cI*857, was created. (Elvin, C. M., *et al.*, Modified bacteriophage lambda promoter vectors for overproduction of proteins in *Escherichia coli*, *Gene* 87, 123–126 (1990)). This protein folds normally and functions as a repressor at 30°C, but unfolds and cannot function when the temperature is raised to 42°C. Therefore, transcription can be induced from a P_L or P_R promoter in cells expressing *cI*857 when the temperature is raised to 42°C. The *cI*857 gene is often included on the P_L or P_R expression plasmid. Because the P_L and P_R promoters are strong promoters when induced, but the temperature-dependent regulation of the cI857 protein is not ideal for many studies, variants of the P_L and P_R promoters have been made where the lambda repressor binding sites are replaced by the binding site for a different repressible transcription factor (see the example in the *TetR* promoter section below).

TetR promoter

The Tn10 transposable element from *E. coli* contains genes that confer tetracycline resistance. Tetracycline is an antibiotic that kills bacteria by binding to the bacterial ribosome and inhibiting translation. A bidirectional promoter between the *tetR* ORF and the *tetA* ORF regulates the transcription of both of these genes. This region contains two *tetR* promoters and one *tetA* promoter. In the absence of TetR protein, both of these genes are transcribed. The TetA protein pumps tetracycline out of the cell. The TetR protein forms dimers that bind to two sites in the *tetR/tetA* promoter and inhibit the transcription of both of these genes. In the presence of tetracycline, the TetR protein binds tetracycline and does not bind to the promoter, thus allowing transcription of the *tetR* and *tetA* genes. The *tetA* promoter is more strongly repressed when off and more strongly expressed when on compared to either *tetR* promoter.

Transcription of the *tetR* and *tetA* genes in the presence of tetracycline

The same DNA sequence between the *tetR* and *tetA* ORFs contains 3 different overlapping promoters. The 5' to 3' (top) strand of DNA is shown for the P_{R1} and P_{R2} promoters, but the P_A promoter is transcribed in the opposite direction and the 3' to 5' (bottom) strand of DNA is shown. The -35 element, -10 element, and transcription start site are shown in bold and underlined for each promoter. In the absence of tetracycline, the binding of TetR protein dimers to the same TetR binding sites (dashed rectangles) represses transcription from all three promoters. TetR bound to tetracycline cannot bind the promoters.

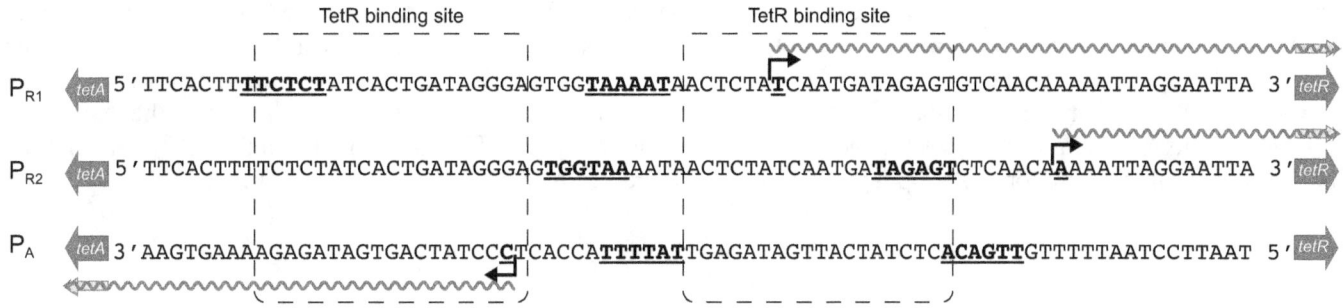

Hillen, W. & Berens, C. Mechanisms underlying expression of Tn10 encoded tetracycline resistance. *Annual Review of Microbiology* 48, 345–69 (1994).

In engineering, many "*tetR*" promoters are artificial promoters that are variations of other promoters that have been modified to include TetR-binding sites (operators) so they can be repressed by TetR. For example, the lambda promoter P_L was modified by replacing lambda repressor binding sites with TetR binding sites, creating a promoter ($P_{LtetO-1}$) whose expression can be regulated over a 5000-fold range using tetracycline, a much greater range than for the other promoters described here. This promoter can also be used in cells that do not produce TetR protein, so that it is effectively on constitutively at a high level.

Replacing the lambda repressor binding sites in lambda P_L with TetR binding sites allows for tetracycline induction

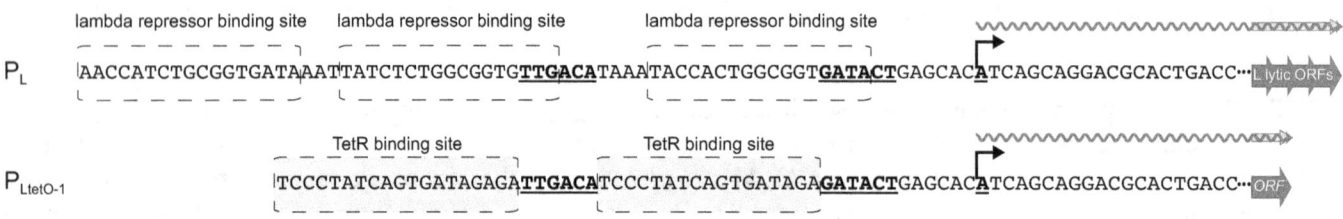

Lutz, R. & Bujard, H. Independent and tight regulation of transcriptional units in *Escherichia coli* via the LacR/O, the TetR/O and AraC/I$_1$-I$_2$ regulatory elements. *Nucleic Acids Research* 25, 1203–1210 (1997).

Sometimes multiple TetR binding sites may be linked together to form a "tetracycline responsive element" (TRE) that can be linked to a minimal promoter sequence to create an artificial kind of tetracycline-responsive promoter. TetR has also been altered and fused to other proteins to allow for novel actions at TREs. Note that the bacterial TetR system is described here. This system has also been modified to allow for inducible or repressible gene expression in eukaryotic cells because tetracycline is able to enter eukaryotic cells and very low amounts are needed to affect the binding of TetR.

T7 Promoter

The bacteriophage T7 promoter is used for engineering *E. coli*, but it cannot be transcribed by a normal *E. coli* RNA polymerase because it does not contain sequences, such as the -10 and -35 elements, that are recognized by *E. coli* RNA polymerase:

T7 promoter: 5' TAATACGACTCACTATA**G** 3'   (The final "G" is the first transcribed nucleotide)

Instead, bacteriophage T7 RNA polymerase is required to transcribe from the T7 promoter. T7 RNA polymerase is capable of transcribing longer RNA molecules than *E. coli* RNA polymerase.

The T7 promoter is often used *in vivo* or *in vitro* for the transcription of RNA molecules that are intended to be used or studied as RNA instead of being translated into protein. However, the T7 promoter can also be used for the expression of mRNA molecules for the purpose of producing proteins. The T7 promoter allows for the transcription of RNA molecules at such high levels that bacteria expressing a gene from a T7 promoter can be killed due to the depletion of resources. Therefore, expression from a T7 promoter in *E. coli* is usually regulated for research purposes. This is often accomplished by regulating the expression of the gene that encodes T7 RNA polymerase. One method is to use an IPTG-inducible *lac* promoter to regulate the transcription of the T7 RNA polymerase ORF. Induction of the expression of T7 RNA polymerase in turn causes the induction of transcription from a T7 promoter. (Davanloo, P., *et al*. Cloning and expression of the gene for bacteriophage T7 RNA polymerase. *PNAS* 81, 2035-2039 (1984)).

Appendix 4: Restriction Enzymes

The New England Biolabs Catalog (https://www.neb.com/) is recommended for a comprehensive listing of restriction enzymes, as well as useful lists of isoschizomers (enzymes that recognize and cut the same sites), compatible cohesive ends, sensitivity to methylation, and required numbers of extra base pairs of DNA surrounding recognition sites. A small selection of restriction enzymes is listed here.

Restriction Enzyme	Recognition and Cut Sites	Source species	Methylation sensitivity (bacteria)	Methylation sensitivity (mammals)	Type
*Age*I	A▼CCGGT TGGCC▲A	*Agrobacterium gelatinovorum*		^{m}CpG	II
*Apa*I	GGGCC▼C C▲CCGGG	*Acetobacter pasteurianus*	dcm^+ (C^mCWGG)	^{m}CpG	II
*Bam*HI	G▼GATCC CCTAG▲G	*Bacillus amyloliquefaciens* H			II
*Bbs*I	GAAGAC(N)$_2$▼ CTTCTG(N)$_6$▲	*Brevibacillus laterosporus*			IIS
*Bcl*I	T▼GATCA ACTAG▲T	*Bacillus caldolyticus*	dam^+ (G^mATC)		II
*Bgl*II	A▼GATCT TCTAG▲A	*Bacillus globigii*			II
*Bsa*I	GGTCTC(N)$_1$▼ CCAGAG(N)$_5$▲	*Bacillus stearothermophilus* 6-55	dcm^+ (C^mCWGG)	^{m}CpG	IIS
*Bsa*XI	▼$_9$(N)ACNNNNNCTCC(N)$_{10}$▼ ▲$_{12}$(N)TGNNNNNGAGG(N)$_7$▲	*Bacillus stearothermophilus* 25B			IIS
*Bsm*BI	CGTCTC(N)$_1$▼ GCAGAG(N)$_5$▲	*Bacillus stearothermophilus* B61		^{m}CpG	IIS
*Bst*EII	G▼GTNACC CCANTG▲G	*Bacillus stearothermophilus* ET			II
*Dpn*I	CH$_3$ | GA▼TC CT▲AG | CH$_3$	*Diplococcus pneumoniae* G41	[Only cuts dam^+ (G^mATC) methylated]	^{m}CpG	II

Restriction Enzyme	Recognition and Cut Sites	Source species	Methylation sensitivity (bacteria)	Methylation sensitivity (mammals)	Type
*Dpn*II	▾GATC CTAG▴	*Diplococcus pneumoniae* G41	dam⁺ (G^{m}ATC)		II
*Eco*RI	▾GAATTC CTTAAG▴	*E. coli* RY13		^{m}CpG	II
*Eco*RV	GAT▾ATC CTA▴TAG	*E. coli* strain		^{m}CpG	II
*Fok*I	GGATG(N)₉▾ CCTAC(N)₁₃▴	*Flavobacterium okeanokoites*	dcm⁺ (C^{m}CWGG)	^{m}CpG	IIS
*Hin*dIII	▾AAGCTT TTCGAA▴	*Haemophilus influenzae* Rd			II
*Hpa*II	▾CCGG GGCC▴	*Haemophilus parainfluenzae*		^{m}CpG	II
*Kpn*I	GGTAC▾C C▴CATGG	*Klebsiella pneumoniae*			II
*Mfe*I	▾CAATTG GTTAAC▴	*Mycoplasma fermentans*			II
*Msp*I	▾CCGG GGCC▴	*Moraxella* species			II
*Nde*I	CA▾TATG GTAT▴AC	*Neisseria denitrificans*			II
*Nhe*I	▾GCTAGC CGATCG▴	*Neisseria mucosa heidelbergensis*		^{m}CpG	II
*Not*I	▾GCGGCCGC CGCCGGCG▴	*Nocardia otitidis-caviarum*		^{m}CpG	II
*Pst*I	CTGCA▾G G▴ACGTC	*Providencia stuartii* 164			II
*Pvu*I	CGAT▾CG GC▴TAGC	*Proteus vulgaris*		^{m}CpG	II
*Pvu*II	CAG▾CTG GTC▴GAC	*Proteus vulgaris*			II

Restriction Enzyme	Recognition and Cut Sites	Source species	Methylation sensitivity (bacteria)	Methylation sensitivity (mammals)	Type
*Sac*I	GAGCT▼C C▲TCGAG	*Streptomyces achromogenes*			II
*Sal*I	G▼TCGAC CAGCT▲G	*Streptomyces albus* G		^{m}CpG	II
*Sca*I	AGT▼ACT TCA▲TGA	*Streptomyces caespitosus*			II
*Sma*I	CCC▼GGG GGG▲CCC	*Serratia marcescens*		^{m}CpG	II
*Spe*I	A▼CTAGT TGATC▲A	*Sphaerotilus* species			II
*Xba*I	T▼CTAGA AGATC▲T	*Xanthomonas badrii*	dam$^+$ (G^{m}ATC)		II
*Xho*I	C▼TCGAG GAGCT▲C	*Xanthomonas holcicola*		^{m}CpG	II
*Xma*I	C▼CCGGG GGGCC▲C	*Xanthomonas malvacearum*		^{m}CpG	II

How restriction enzymes are named:
The first 3 letters in the name are an abbreviation of the genus and species name of the source bacteria. This portion of the name is italicized because genus and species names are italicized. There is sometimes a letter next that indicates the particular strain of that species of bacteria. The name ends with a capitalized roman numeral that indicates which enzyme number it was among the enzymes found in those bacteria.

Methylation sensitivity:
The *E. coli* Dam methyltransferase adds a methyl group to the A in the sequence G^{m}ATC. The *E. coli* Dcm methyltransferase adds a methyl group to the C in the sequence C^{m}CWGG where "W" is an A or T. Human methyltransferase will methylate the C in a subset of CpG dinucleotides. Some restriction enzymes will not cut if a nucleotide in their site is methylated and these enzymes are marked above. If you need to use an enzyme that is sensitive to Dam or Dcm methylation, you can isolate your plasmid DNA from an *E.coli* strain that is *dam*$^-$ and *dcm*$^-$. *Dpn*I is different from the other listed enzymes in that it specifically cuts Dam-methylated DNA and will not cut unmethylated DNA.

Restriction enzyme types:
There are at least four classes of restriction enzymes. The most commonly used kind in research are type II restriction enzymes. The most commonly used group of type II enzymes cut at specific positions in their recognition sites. Most of these have palindromic, symmetrical recognition sites, but some have asymmetric recognition sites. Type IIS enzymes are a different subset of type II enzymes that cut close to, but not within their recognition sites and they have asymmetric recognition sites.

Appendix 5: Reference Information

Relative Electronegativities

2.1 H																	
1.0 Li	1.5 Be											2.0 B	2.5 C	3.0 N	3.5 O	4.0 F	
1.0 Na	1.2 Mg											1.5 Al	1.8 Si	2.1 P	2.5 S	3.0 Cl	
0.9 K	1.0 Ca	1.3 Sc	1.4 Ti	1.5 V	1.6 Cr	1.6 Mn	1.7 Fe	1.7 Co	1.8 Ni	1.8 Cu	1.6 Zn	1.7 Ga	1.9 Ge	2.1 As	2.4 Se	2.8 Br	
0.9 Rb	1.0 Sr	1.2 Y	1.3 Zr	1.5 Nb	1.6 Mo	1.7 Tc	1.8 Ru	1.8 Rh	1.8 Pd	1.6 Ag	1.6 Cd	1.6 In	1.8 Sn	1.9 Sb	2.1 Te	2.5 I	
0.8 Cs	1.0 Ba	1.1 La	1.3 Hf	1.4 Ta	1.5 W	1.7 Re	1.9 Os	1.9 Ir	1.8 Pt	1.9 Au	1.7 Hg	1.6 Tl	1.7 Pb	1.8 Bi	1.9 Po	2.1 At	
0.8 Fr	1.0 Ra	1.1 Ac															

Difference in Electronegativities

0.0 — 0.4 nonpolar | 0.4 — 1.7 polar | 1.7 — 4.0 ionic

Functional group	General structure	Structure in cell (aqueous, pH 7)
hydroxyl	~C–OH	~C–OH
carbonyl	~C=O	~C=O
carboxylic acid	~C(=O)OH	~C(=O)O⁻
amino	~NH$_2$	~NH$_3^+$
phosphate	~O–P(=O)(OH)–OH	~O–P(=O)(O⁻)–O⁻

$R = 8.314$ J/mol·K

Avagadro's number: 6.02×10^{23}

1 M = 1 mol/L

A 1% solution (weight/volume) contains 1 g of solute per 100 mL of solution

Equations

$\Delta G° = \Delta H° - T\Delta S°$

If: $aA + bB \rightleftharpoons cC + dD$

Then: $\Delta G = \Delta G° + RT \ln \left[\dfrac{[C]^c[D]^d}{[A]^a[B]^b}\right]$

$K_{eq} = \dfrac{[C]^c[D]^d}{[A]^a[B]^b}$

$pH = -\log[H^+]$

$K_w = [H^+][OH^-] = 1 \times 10^{-14}$

$[\text{stock}] \times V_{(\text{stock})} = [\text{diluted}] \times V_{(\text{diluted})}$
$M_1V_1 = M_2V_2$

T_m of a primer $= (4°C)(\text{\# of Cs or Gs}) + (2°C)(\text{\# of As or Ts})$

Constants and Conversions

Millimeters: 1×10^3 mm = 1 m
Micrometers: 1×10^6 μm = 1 m
Nanometers: 1×10^9 nm = 1 m
Angstroms: 1×10^{10} Å = 1 m
Picometers: 1×10^{12} pm = 1 m

1000 bp = 1 kb
Kelvin = °C + 273.15

1st:	2nd:	U		C		A		G		3rd:
U		UUU UUC	Phe	UCU UCC UCA UCG	Ser	UAU UAC	Tyr	UGU UGC	Cys	U C
		UUA UUG	Leu			UAA UAG	Stop Stop	UGA UGG	Stop Trp	A G
C		CUU CUC CUA CUG	Leu	CCU CCC CCA CCG	Pro	CAU CAC	His	CGU CGC CGA CGG	Arg	U C A G
						CAA CAG	Gln			
A		AUU AUC AUA	Ile	ACU ACC ACA ACG	Thr	AAU AAC	Asn	AGU AGC	Ser	U C
		AUG	Met			AAA AAG	Lys	AGA AGG	Arg	A G
G		GUU GUC GUA GUG	Val	GCU GCC GCA GCG	Ala	GAU GAC	Asp	GGU GGC GGA GGG	Gly	U C A G
						GAA GAG	Glu			

The Genetic Code

This table shows amino acids specified by mRNA codons. The AUG start codon and the three stop codons are highlighted.

Amino Acids:

Ala **A** Alanine	**Arg** **R** Arginine	**Asn** **N** Asparagine	**Asp** **D** Aspartic acid
Cys **C** Cysteine	**Glu** **E** Glutamic acid	**Gln** **Q** Glutamine	**Gly** **G** Glycine
His **H** Histidine	**Ile** **I** Isoleucine	**Leu** **L** Leucine	**Lys** **K** Lysine
Met **M** Methionine	**Phe** **F** Phenylalanine	**Pro** **P** Proline	**Ser** **S** Serine
Thr **T** Threonine	**Trp** **W** Tryptophan	**Tyr** **Y** Tyrosine	**Val** **V** Valine

Appendix 5: Reference Information

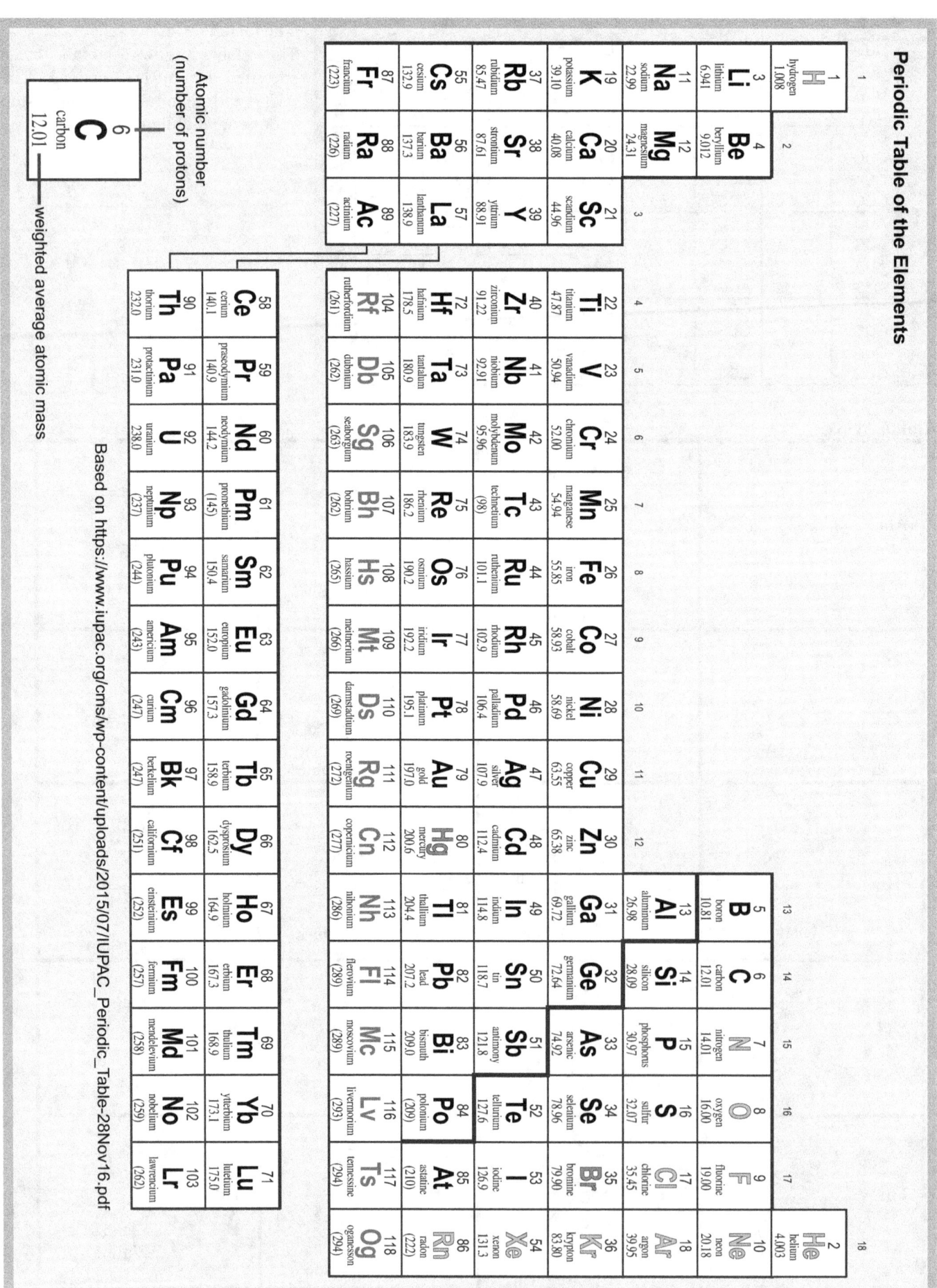

www.ingramcontent.com/pod-product-compliance
Lightning Source LLC
Chambersburg PA
CBHW080914170526
45158CB00008B/2108